Methods of Estimating the Volume of Undiscovered Oil and Gas Resources

Methods of Estimating the Volume of Undiscovered Oil and Gas Resources

edited by
John D. Haun

Published by
The American Association of Petroleum Geologists
Tulsa, Oklahoma, U.S.A., 1975

Contents

Preface

AAPG research conferences were initiated to promote the advancement of knowledge relating to petroleum geology and allied subjects. This volume, the first in a series, has been published as quickly as possible following the conference at Stanford University on "Methods of Estimating the Volume of Undiscovered Oil and Gas Resources." The subject is important and timely, especially in view of developing energy shortages in various parts of the world.

Many research conferences have a rule against publication of presented papers or discussions. This is thought to encourage free exchange of ideas and presentation of partially completed projects so that participants may compare their methods and objectives with those of their colleagues. The philosophy behind AAPG research conferences, however, is a modification of this "no publication" concept in an effort to make as much information as possible available to the membership and to the public. If a significant number of the participants agree to the publication of their papers, as is the case with the Stanford conference, then an effort will be made to have them published. One of the participants did not wish to have his work published at present, and another was unable to meet the deadline for submittal of manuscripts.

We have tried to make this volume as complete a record of the conference as possible, but it should be obvious that this is a progress report on efforts to perfect methods of estimating the petroleum potential of areas which have varied levels of exploration maturity. Some of the methods being used today were not presented at the conference.

JOHN D. HAUN
Golden, Colorado
January 31, 1975

Methods of Estimating the Volume of Undiscovered Oil and Gas Resources—
AAPG Research Conference[1]

JOHN D. HAUN[2]

Shortages of gasoline, heating oil, and natural gas and concern for the increasing usage of the world's depletable energy resources have resulted in a proliferation of reserve and resource estimates that almost defy rational analysis. Two recently published estimates of "undiscovered" oil and gas resources of the United States illustrate the problem: (1) the U.S. Geological Survey optimistically estimated (U.S. Dept. of Interior News Release, March 26, 1974) that the United States contains 200 - 400 billion bbl of oil and 1,000 - 2,000 Tcf of natural gas yet to be discovered; (2) Mobil Oil Corporation's "best" estimate is 88 billion bbl of oil and 443 Tcf of natural gas yet to be discovered in the United States (Gillette, 1974). The pessimistic estimate by Mobil probably represents the view of a majority of major-oil-company exploration personnel. Which of these estimates is correct—if either—is important not only to the petroleum industry but also to the various governmental agencies now conducting long-range energy-resource planning.

In an effort to compare, and perhaps to refine, methods of estimating oil and gas resources, a research conference sponsored by The American Association of Petroleum Geologists (AAPG) was convened at Stanford University August 21 - 23, 1974. The conference was attended by 50 government, university, and industry geologists—from the United States, Canada, and France—who have been concerned with petroleum resource estimates for several years (and who are responsible for many of the published estimates). Kenneth H. Crandall was convener of the conference, and John W. Harbaugh arranged the housing, food, and meeting facilities for the participants. It was an interesting, productive, and well-organized conference, and the participants agreed that a similar meeting should be held in about 2 years.

Crandall made a plea for clarity in definition of resource categories as a means of increasing meaningful communications among geologists and between geologists and the public (or the government). He pointed out that the purpose of the conference was to gain understanding, but not necessarily uniformity, of methods of resource estimation.

In an introductory commentary, Michel T. Halbouty reviewed the policy implications of pessimistic versus optimistic resource estimates. If it is true that we have already discovered more than half of the producible oil and gas, the remaining life of the domestic exploration industry (especially the onshore part) will be short and increased prices will not stem the downward trend in rate of discovery. If, on the other hand, the undiscovered petroleum resources are as vast as some estimates indicate, the domestic industry should have a life expectancy well into the next century and, with adequate economic incentives, we should be able to maintain a secure petroleum inventory during the developmental stages of alternate energy sources. Methods of resource assessment must incorporate the impact of new exploratory tools and other technological factors, the change from structural-trap to stratigraphic-trap exploration, and the existence of large unexplored areas. In addition, long-

[1] Manuscript received, January 31, 1975.
[2] Colorado School of Mines, Golden, Colorado 80401.

range predictions "...must be made with the realization that low estimates could discourage future investment or high estimates could mislead,
and ultimately disappoint, the American public."

Definitions of reserves and resources were discussed by V. E. McKelvey.
"Reserves are considered to be identified deposits recoverable under
existing economic and technologic conditions. Resources include undiscovered deposits of the same quality as reserves, as well as deposits
presently unrecoverable for either economic or technologic reasons."
These broad definitions and several subcategories of reserves and resources have been adopted recently by the U.S. Geological Survey and the
U.S. Bureau of Mines. Differences in definitions of resource subdivisions
(especially "potential supply") account for some of the differences in
estimates among various conference participants. McKelvey emphasized the
hierarchical relationship of subcategories, in both "identified" and
"undiscovered" resources, to the degree of economic feasibility and the
degree of geologic assurance.

Harry L. Thomsen described the recently formed Resource Appraisal
Group (RAG) of the U.S. Geological Survey. The purpose of the Resource
Appraisal Group, which will consist of 12-15 professionals, is to "esta-
blish a coordinated continuous system" for oil and gas resource assessment, on a geologic-province basis. During the next 3 years, RAG, in
cooperation with AAPG, will make statistical analyses based on computerized
data banks and will analyze the important petroleum provinces of the
world. Reporting will be in well-defined categories, and ranges of estimates will indicate degrees of uncertainty. Establishment of RAG was
necessitated by the increasing pressure from governmental bodies for
more accurate resource estimates.

The method of resource estimation used by Thomas A. Hendricks involves
(1) computing the quantity of oil and gas *in place* in intensively explored
areas, (2) dividing these areas "into a graded series of categories,"
(3) assigning inadequately explored areas to similar categories (based
on results of past exploration and geologic characteristics), and (4)
multiplying each inadequately explored area (in 1,000-sq mi increments)
by a "...factor established for the geologically similar explored area."
The difficulty with the method, as Hendricks pointed out, is the subjective determination of "adequately explored areas" and "geologic factors"
considered critical for comparative purposes.

William W. Mallory described the purpose of the "Accelerated National
Oil and Gas Resource Evaluation" (ANOGRE) project. The method is a slight
variation of the Hendricks method in that it involves rock *volume* rather
than *area* (although the volume of potentially petroleum-productive rock
is important in the Hendricks method also). The conterminous United
States is divided into 68 provinces and 30 major stratigraphic units
(MSU's). A dry hole is considered to condemn a volume of rock equal to
that occupied by the average pool in the province and MSU being analyzed.
Formulation of the ANOGRE method may be stated as:

$$\frac{V_{drilled}}{HC_{known}} = \frac{V_{potential}}{HC_{unknown}} \quad (f),$$

where $V_{drilled}$ is the volume of rock tested by development wells in known
pools plus the volume of rock drilled and found barren; $V_{potential}$ is the
volume of rock which seems capable of producing but has not been drilled;
HC_{known} is the volume of hydrocarbons discovered; $HC_{unknown}$ is the computed volume of hydrocarbons yet to be found; and f is a numerical richness factor.

A lively discussion ensued regarding the logic of assigning values of 1.0 and 0.5 to the probability factor, f. It was generally thought that f was smaller than 0.5. Peter Rose promised that the ANOGRE group would take another look at the value of f, and such studies are now under way. It was also argued that the volume condemned by a dry hole should be considerably larger than the *average* pool because *average* pools account for only a small part of the oil discovered—giant pools account for 80 percent of the oil, and a much smaller number of dry holes would be sufficient to indicate that a province (or MSU) could not possibly contain as large a volume of oil as the province (or MSU) with which it is being compared. Several participants urged the Survey to consider incorporating in the ANOGRE system probability ranges and "most likely" values established by Monte Carlo simulation. There was general consensus that several methods of assessment were necessary to verify the accuracy or reality of estimates.

Lewis G. Weeks stated that nothing is accomplished "...by introducing an extensive array of vaguely defined, questionably contributory, and time-consuming subcategory terminologies," and he described 29 factors "...that control the incidence of oil occurrence." Among the factors that should be considered are depositional basin shape, rate of filling, tectonic position, physical and chemical environments, depositional and stratigraphic characteristics, evidences of petroleum, and exploration history. Weeks reviewed world energy demand and predicted that world petroleum production will reach a peak near the year 2000.

Methods of estimating the petroleum resource potential of Canada were reviewed by J. W. Porter and R. G. McCrossan. Areas (basins) of the world were analyzed and classified on the basis of "evolution in basin styles" and "distinctive families of trapping configurations" that have distinctive ranges of hydrocarbon potential. Details of their assessment methods have been published by the Canadian Society of Petroleum Geologists (McCrossan and Porter, 1973). The comparison of undrilled areas with presently petroleum-productive areas, in terms of depositional history and crustal position, is similar to the methods advocated by other speakers. Questions were raised as to the adequacy of the data base for the world's sedimentary basins. The suggestion was made that a cooperative effort in basin analysis be made with the United States.

Ian H. Mackay and F. K. North advocated the use of probability factors in analyses of petroleum potential. They pointed out that in recent years there has been a precipitous decline in discovery rate, especially in the discovery of giant and supergiant fields (the only ones that make any appreciable impact on the world's petroleum supply). They predicted that most of the large fields remaining to be discovered will be in offshore structural traps "in remote, hazardous, or insecure regions." The probabilities that a basin contains a 500-million bbl field, a 1-billion bbl field, or more than one supergiant field are small. The authors concluded "...that the Free World's undiscovered reserves of conventional oil are smaller than the amount already produced." Their use of the non sequitur "undiscovered reserves" was questioned (by definition, "reserves" have been *discovered!*).

Several participants voiced opposition to the pessimistic view of Mackay and North. D. W. Axford pointed out that in recent years the largest gas fields in Canada have been discovered in the Arctic Islands. B. W. Beebe cited Jay field in Florida, Gomez in Texas, and Brady in Wyoming as indicative of the large fields that *are* being found in the United States. McKelvey reminded us that economic and political stimuli affect projections on a time frame and, therefore, time projections cannot tell us what is left in the ground; on the other hand, an exponential

growth is untenable and leads to an unrealistic world. A million-year
supply of *anything*, at today's usage rate, will be exhausted in 343 years
at a 3-percent annual rate of increasing consumption.

A system of basin analysis using a computerized analogy technique
was described by Christian Bois. In the selection of widely accepted
and understood "units" for comparison, there are two concerns: (1) a
reference basin (or area) *must* be completely explored, and (2) large
basin (or area) segments that contain several habitats of petroleum occurrence and trap types must not be used. To avoid the difficulty in characterizing large areal units, basins are subdivided into "zones" of related
groups of fields that have similar age, trap type, source rock, reservoir
rock, petroleum quality, etc. There are approximately 50 geologic and
50 hydrocarbon numerical parameters that form a logical progression, or
hierarchy, of qualities. Comparison of similarity or lack of similarity
is made among the zones by the use of cluster analysis. Classes are
designated on the basis of the number of common parameters.

The membership, organization, and work of the Potential Gas Committee
(PGC) were reviewed by B. W. Beebe, R. J. Murdy, and E. A. Rassinier.
They emphasized the fact that a common method of estimating undiscovered
natural gas resources, by first estimating oil resources and then deriving
gas resources on the basis of a presumed gas/oil ratio, leads to spurious
conclusions. Gas resources should be analyzed separately; gas commonly
has a different habitat than oil. The PGC estimates the quantity of
undiscovered natural gas in each geologic province by a geologic-volumetric, or "attribution," technique that is based on the quantity of gas
per unit of reservoir volume, derived from the volume of individual zones
that are productive, condemned, or potentially productive. The estimates
are classified, by decreasing degrees of certainty, into "probable,"
"possible," and "speculative" categories. Techniques used in making
estimates are described in the most recent PGC report (Potential Gas
Committee, 1973). PGC estimates are widely quoted and used by governmental agencies and the petroleum industry.

Several conference participants strongly suggested that more complete
PGC data be published, perhaps in tabular form, to show the details on
which PGC estimates are based. Estimates based on confidential data not
available to the public in the usual scientific manner, and therefore
difficult to verify, are subject to suspicion. This same negative com-
ment applies also to unsupported (undocumented) recently published estimates of oil and gas resources by major petroleum companies.

Earl Cook contributed to the pessimistic view with his summary of
published estimates of undiscovered oil resources made within the past
10 years (some estimates are 15 times larger than others). He concluded
that the range of estimates renders them useless in formulation of national
strategy. Estimates "...tend to project past costs of exploitation and
to ignore exponential increases of work cost with depth," and "...they
also ignore the probability that 'substitution' technology will outpace
petroleum technology....It may be that nongeologic methods of estimating
...are better guides to national policy than are geologic methods"—because the time required for making estimates is important to the nation.

According to R. H. Nanz (manuscript not submitted), there is "...no
way at this time to estimate with any useful degree of accuracy the
volume of undiscovered oil and gas in a given unexplored province"; but
he described qualitative methods used to assess the "reasonable maximum
petroleum potential of a province or basin." Analyses are made of critical
conditions such as the availability of hydrocarbons (factors considered
include source beds, maturation, migration paths), presence of reservoirs
(considering such things as cementation, compaction, solution), presence
of sealing beds, presence of traps and trap capacity, timing, paleo-

structure, and comparison with productive regions. The range of recoverable volume that might be present is called the "scope volume" (a term suggested by A. W. Bally). More direct methods are used also—marine hydrocarbon surveys, sidescan sonar, and seismic reflection amplitudes (for gas accumulations). Remaining questions relate to prospect definition, adequacy of operational technology, and economic feasibility.

Nanz said that Shell had analyzed the sedimentary basins of the world in an effort to find a relation between basin type and content of recoverable petroleum. They "...have not succeeded in developing any systematic relation between recoverable hydrocarbon and basin thickness, areal extent, or sediment volume." The expectation for an unexplored basin, therefore, "...is in the range of zero to some number probably less than our reasonable maximum volume." The estimated range should be more accurate, however, if 20 or more unexplored basins are considered for their "total expectation volume." On this basis, Shell estimates that undiscovered oil in the United States is in the order of 100-125 billion bbl (20 billion bbl in the onshore conterminous United States). Exploration on public lands should not be delayed pending more accurate estimates.

Some of the possible uses of large data banks were described by John L. Stout. More than 755,000 well histories and production data for three fourths of the petroleum production in the United States are available in the computerized data banks of Petroleum Information Corporation. Assessments of undiscovered petroleum must rely on accurate knowledge of the results of past exploration: cumulative production, oil and gas volumes initially present, success ratios, economic data, etc. Projections for the immediate future should be most accurate; systematic updating could maintain the projections as a periodic series.

A probabilistic model of the petroleum discovery process was described by Gordon M. Kaufman, Y. Balcer, and D. Kruyt. Statistical regularities (lognormal distributions) in the number and size distribution of discovered pools perhaps can be used as a model for predicting the distribution of undiscovered pools. This method involves the probabilistic characterization of the historical way in which a "play" develops (the largest pool is usually discovered early in the play). The probability that the "next" discovery will be of size "A" is the ratio of "A" pools to the sum of sizes of undiscovered pools (sampling without replacement). Analyses of the major Alberta plays were used as an empirical validation of the predictive character of the model.

A group of Exxon geologists (David A. White, Ralph W. Garrett, Jr., G. R. Marsh, R. A. Baker, and H. M. Gehman) presented a series of comprehensive assessment methods that may be used in estimating regional oil and gas potentials. The "probable" supply is estimated by projecting the growth of reserves in known fields on the basis of year-to-year revision ratios (Marsh, 1971). The "possible" supply, in maturely explored formations with declining discovery rates, "...is estimated from the extrapolation of historical discovery rates...expressed as equivalent barrels of oil found per foot of wildcat drilling." It is necessary to estimate the minimum, maximum, and most likely quantities to be discovered per foot drilled, and to estimate the exploratory-well footage that will be drilled (considering economic factors).

The "speculative" supply in unexplored areas or in deeper unexplored parts of the stratigraphic section is estimated by geologic analysis and comparison. Minimum, maximum, and most likely probabilities are calculated for estimated reservoir volumes and quantity of petroleum per unit volume. All multiple probability additions are made by Monte Carlo simulation. A probability curve for each estimate "...reflects the combined uncertainties in the input factors." An assessment of the undiscovered gas potential of onshore south Louisiana was presented as an illustration of the Exxon group's method.

Hollis D. Hedberg submitted two short papers relating to the precision of resource estimates and the "volume-of-sediment fallacy" in resource estimates. He advocated the use of round numbers or probability ranges from zero to "X" barrels for estimates of petroleum resources in undrilled areas (precise numbers give an impression of greater accuracy than is warranted). He, too, made the point that it is not possible to select with assurance a satisfactory figure for productivity per unit volume for application to an unknown area.

The petroleum potential of deep ocean areas was reviewed by K. O. Emery. He suggested that primarily marine deposits of continental shelves may contain larger petroleum resources per unit area than land areas, but lack of drilling makes estimates unreliable. Estimates of petroleum potential decrease in reliability from marginal basins to continental slopes and rises, and to the deep ocean floor. JOIDES drilling has intentionally avoided areas of petroleum potential to prevent possible pollution. Therefore, because of the lack of data "...quantitative estimates of undiscovered oil and gas resources of the deep ocean floor are meaningless."

An example of the analytical technique known as the "Kansas Oil Exploration (KOX) Decision System," being developed by the Kansas Geological Survey, was presented by William W. Hambleton, John C. Davis, and J. H. Doveton. A "causal predictive model," in which estimates are derived from such factors as "...geometry and magnitude of source beds and potential reservoir formations," and an "empirical predictive model," which projects exploration history into the future, are incorporated in the KOX system. The apparent geologic character of an area changes during the exploration history. The statistical "experience" of a maturely developed area of central Kansas was applied to a partially developed area in northwestern Kansas. Possible exploration outcomes "...are expressed as conditional probabilities drawn from contingency tables relating known production to perceived subsurface geology." These authors used "unknown reserves" as a substitute for "undiscovered resources"!

An abstract submitted by Harrison H. Schmitt for the Conference (not submitted for publication) contained the estimate that an assessment of all unexplored domestic areas could be carried out at a cost of approximately $2 billion. This estimate was one of the conclusions of a committee that met December 7, 1973, at the California Institute of Technology. Methods of conducting this assessment were not mentioned.

Robert W. Jones and J. R. Baroffio (see paper in this volume by Jones) argued that, inasmuch as every basin is unique, there is such a lack of correlation between geologic parameters and petroleum resources (per cubic mile) in well-explored basins that this comparative technique is unsatisfactory. Jones proposes an analytical method of resource estimation based on "...four equally important factors: reservoir (R), trap (T), source (S), and migration (M)." The factoring model may be stated as: Estimated reserves/cu mi = $R \times T \times S \times M$, where R is the fraction of a basin which could contain producible petroleum, T is the fraction of R which is in trap position, S is the ratio of petroleum in the source rock to trap capacity (RT), and M is the ratio of petroleum in traps to petroleum in the source rock.

R, T, S, and M are rated on a normalized 0-10 scale. "The 10 rating was empirically determined from the maximum value observed in over 50 well-explored basins." Jones believes that accuracy in rating decreases from T to R to S to M, and that increased knowledge of S and M could lead to more accuracy of prediction. Again, it was pointed out that the word "reserves" was used to mean undiscovered "resources."

Essentially nongeologic, time-dependent statistical methods of estimating undiscovered domestic or world petroleum potential were not discussed in detail at the conference, but several references to the published estimates of M. King Hubbert (not present) were made. During discussions, C. L. Moore briefly described the fundamental differences between his estimates, based on Gompertz-curve projections, and Hubbert's estimates, based on logistic-curve projections. Several participants argued that time-dependent projections do not account adequately for economic, political, and technologic changes.

No formal conclusions or recommendations were adopted by the conferees, but there was general consensus on the following points:

1. A more uniform series of well-defined terms to designate categories of petroleum resources should be adopted as a means of increasing understanding and communication among geologists, engineers, and various segments of the public. Resource classification and definition of terms recently adopted jointly by the U.S. Geological Survey and the U.S. Bureau of Mines (U.S. Dept. of Interior, 1974) should be restated (for petroleum) in terms that parallel traditional usage by the petroleum industry.

2. Assumptions, data sources, regional and stratigraphic information, and methods of making resource estimates should be published with the estimates so that scientific credibility will not be in doubt. (Exxon and KOX Project geologists presented analyses based on *public* rather than *proprietary* information.)

3. Methods of assessment should be flexible; no single method is applicable to all provinces, basins, stratigraphic sequences, or stages of exploration effort.

4. More than one method of assessment should be used for all areas so that reliability can be enhanced and reasonableness of estimates can be tested.

5. Estimates should be stated as ranges of probability, and probabilities should be summed by Monte Carlo simulation rather than by simple addition of probability minimums and maximums.

REFERENCES CITED

Gillette, Robert, 1974, Oil and gas resources; did USGS gush too high?: Science, v. 185, no. 4146, July 12, p. 127-130.

Marsh, G. R., 1971, How much oil are we really finding?: Oil and Gas Jour., v. 69, no. 14, April 5, p. 100-104.

McCrossan, R. G., and J. W. Porter, 1973, The geology and petroleum potential of the Canadian sedimentary basins—a synthesis, *in* Future petroleum provinces of Canada—their geology and potential: Canadian Soc. Petroleum Geologists Mem. 1, p. 589-720.

Potential Gas Committee, 1973, Potential supply of natural gas in United States (as of December 31, 1972): Golden, Colorado, Colorado School of Mines Foundation, Inc., 48 p.

U.S. Department of the Interior, 1974, Estimates of reserves of minerals and mineral fuels on federal leases: Federal Register, v. 39, no. 145, July 26, p. 27334-27335.

Methods of Estimating the Volume of Undiscovered Oil and Gas Resources—Introductory Remarks[1]

MICHEL T. HALBOUTY[2]

Because of the current and impending energy shortages, there is a healthy interest in estimates of reserves and resources of oil and gas. However, the controversy about the accuracy of recent estimates of domestic undiscovered petroleum resources unquestionably is retarding the development of a sound national energy policy.

One camp in this controversy maintains, on the basis of past domestic industry performance or presently identifiable "plays" and prospects, that we already have discovered well over half of the producible oil and gas in the United States, and that a little more than half of that already has been produced. The implication of this view is that the remaining life of the domestic exploration industry, particularly onshore, is short—perhaps 10 years. A further implication is that increased prices for petroleum will not result in sizable new discoveries and, therefore, money would be better used for development of alternate energy resources.

The other camp maintains, on the basis of geologic analogy and the existence of large unexplored areas, that the domestic situation is not so dismal; that considerable oil and gas remain to be found by using new tools, human ingenuity, and the stimulation of increased prices; and that a productive domestic exploration industry will continue for at least another 20-30 years. Further implications are that maximum sustained domestic exploration is a key ingredient to the national energy position, and that we must identify and develop our remaining domestic petroleum resources to serve as security while alternate energy sources are being developed during the next 3 decades.

The business of predicting the unknown—i.e., petroleum resource estimation—is important, and there is no doubt that resource predictions by influential geoscientists will affect the developing national energy policy. We must work to develop a technique or techniques for petroleum resource estimates that will give full credence to new exploratory tools and concepts, and to broader and greater efforts in oil exploration. These estimates must take into consideration the future dominance of stratigraphic-trap exploration and the accumulations that have been identified but cannot be extracted because of economic or technologic factors. Economic or subeconomic resources that are yet to be discovered must also be considered.

In brief, we need a method or methods that will take into account the new techniques and higher prices that surely lie ahead. The reasonably accurate long-range predictions thus generated can be publicly defended and justified. However, they must be made with the realization that low estimates could discourage future investment or high estimates could mislead, and ultimately disappoint, the American public. Furthermore, inaccurate predictions will have a deleterious effect on national energy planning.

[1]Manuscript received, February 5, 1975.
[2]Consulting geologist and petroleum engineer, Houston, Texas 77027.

I want to review briefly a few facts that have been overlooked by those in what I call the "pessimistic group"—facts which, I believe, render their predictions of future domestic petroleum discoveries too low.

DOMINANCE OF STRATIGRAPHIC-TRAP EXPLORATION IN FUTURE

Prediction by analogy works only if one is dealing with similar situations. Those methods of resource prediction that rely on industry's past performance overlook the fact that previous exploration was dominated by structural-trap prospecting. As we move into a period that emphasizes stratigraphic-trap exploration, performance patterns surely will change. Also, the guidelines which govern exploration thinking and procedures frequently have unanticipated results. Were this not true, many of the world's great stratigraphic-trap fields would not have been discovered. Accordingly, we must develop resource appraisal methods which give full weight to the enormous potential of stratigraphic traps.

DEVELOPMENT OF NEW CONCEPTS

The problem with estimating remaining potential on the basis of present concepts is that it presupposes that we now know all the kinds of traps and accumulations that exist. For example, the giant Altamont accumulation in the Uinta basin may represent an entirely "new" kind of accumulation—not structural or stratigraphic but, rather, an accumulation formed by the positioning of rich source rocks and adjacent impervious siltstones and fractured limestones right in the thermal "window" where generation of petroleum occurs. Who would have conceived of such a trap 10 years ago, much less actually explored successfully with that concept in mind?

DEVELOPMENT OF NEW TOOLS

The most obvious example of new exploration tools is the recent advance in seismic techniques that not only permit direct detection of oil and gas accumulations ("bright spots") but also may permit identification on seismic profiles of certain facies changes that could represent stratigraphic traps. When such seismic techniques are improved and adopted throughout industry, stratigraphic-trap prospecting will have come into its own. Other new methods now under investigation include remote sensing and trace-element anomalies caused by microseepage.

EXISTENCE OF LARGE UNEXPLORED OFFSHORE REGIONS

Because very large segments of the United States offshore (e.g., the Atlantic outer continental shelf, the South Texas offshore, the outer Southern California borderland, the Gulf of Alaska, the Arctic Shelf) are essentially undrilled, it is not legitimate to include them within a hypothetical outer-continental-shelf (OCS) total-resource base for purposes of estimations; yet this is precisely what Hubbert (1969) has done. He includes the OCS of all the lower 48 states within his postulated resource base even though most of the United States offshore has, for all practical purposes, been closed to exploration. Hubbert recognizes that Alaska is a "late comer" and adds its potential to that of the onshore and offshore areas of the lower 48 states, but does not concede that large parts of the OCS should be classed as "late comers."

In this connection, in March 1974 the U.S. Geological Survey released its estimates of undiscovered recoverable petroleum resources: 200-400 billion bbl of crude oil and natural gas liquids and 1,000-2,000 Tcf of natural gas (U.S. Dept. of Interior News Release, March 26, 1974).

Based on my previous published efforts and on more than 4 decades of continuous evaluations of our nation's petroleum potential, I am confident that the U.S. Geological Survey's figures are reasonable and acceptable.

One facet of OCS exploration that disturbs me is the large number (195 out of 864) of tracts that have been relinquished and, we might with tongue in cheek use the word "condemned," by one dry hole. I wonder how many of these 195 tracts warranted more exploration—either by geophysical surveying to delineate the anomaly more accurately, or by the drilling of more wells. I also wonder how many of these tracts might have been productive, and how much more reserves would have been added, if exploration and drilling had been pursued.

REFERENCE CITED

Hubbert, King, 1969, Energy resources, *in* Resources and man: San Francisco, W. H. Freeman, p. 157-242.

Concepts of Reserves and Resources [1]

V. E. McKELVEY [2]

ABSTRACT   In the classification of reserves and resources recently adopted by the U.S. Geological Survey and the U.S. Bureau of Mines, several categories of reserves and resources are differentiated which are useful in reporting objectively the state of knowledge about the existence and recoverability of minerals and in establishing targets for exploration and technologic development. In this classification, the terms "reserves" and "resources" are used in the traditional sense. *Reserves* are considered to be identified deposits recoverable under existing economic and technologic conditions. *Resources* include undiscovered deposits of the same quality as reserves, as well as deposits presently unrecoverable for either economic or technologic reasons. These categories are further subdivided to indicate the degree of certainty of their existence and the feasibility of their recovery under present economic conditions. Rather than attempting to predict how much eventually will be found and produced—as in the concept of ultimate reserves or production—estimates in these terms indicate what we know and what we hypothesize or speculate about.

Two subcategories of measured reserves that would be a useful addition to this classification are "developed measured reserves" and "undeveloped measured reserves."

During the past 10 years, I have been developing a classification of mineral reserves and resources that I believe helps to report objectively the status of our knowledge about minable reserves and potential resources. Recently, the U.S. Bureau of Mines and the U.S. Geological Survey have improved this classification and agreed on definitions that will be followed in future reporting. My purpose here is to discuss this classification, the definitions, and the concepts of reserves and resources on which the classification is based.

The classification as agreed on by the Survey and the Bureau of Mines is shown in Figure 1. Most of the terms have a substantial history of use in the mineral industry and, for the most part, they are used here in their traditional sense. The framework for the classification is provided by the two principal variables involved in resource evaluation: (1) knowledge about the existence, quality, and magnitude of individual deposits, and (2) the feasibility of their recovery under existing prices and technology. We recognize two broad categories of deposits with respect to each of these variables—those deposits that are already identified as opposed to those that are undiscovered, and those that are producible now as opposed to those that are presently subeconomic but may become producible with higher prices or advancing technology. In keeping with traditional usage, *reserves* are defined to include only identified deposits presently producible at a profit, and undiscovered and subeconomic material are referred to as *resources*.

In further categorizing identified resources, we have continued to use the terms "measured," "indicated," and "inferred," which were developed by the Survey and the Bureau during World War II as being more suitable for national estimates than the "proved," "probable," and "possible" analogs (three P's) in common use by industry. The terms "proved" and "measured" are essentially synonymous, for both refer to materials so closely sampled that the amount and quality have been established within a relatively small margin of error. The terms "probable" and "possible," however, are not synonymous with "indicated" and "inferred."

[1] Manuscript received, December 30, 1974.
[2] Director, U.S. Geological Survey, Reston, Virginia 22092.

TOTAL RESOURCES

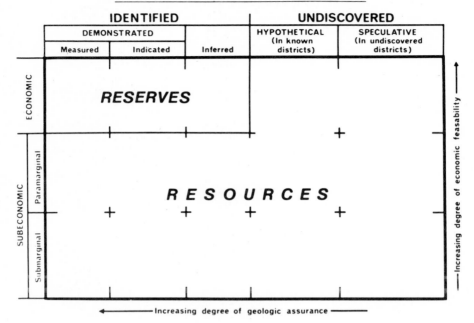

FIG. 1—Classification of mineral reserves and resources approved by U.S. Geological Survey and U.S. Bureau of Mines. Degree of assurance of estimates increases from right to left, and economic feasibility of recovery increases from bottom to top.

"Probable" and "possible" both describe estimates of partly sampled deposits—in some definitions, for example, "probable" is used to describe deposits sampled on two or three sides, and "possible" to describe deposits sampled on only one side. In the Survey and Bureau definition, both types of deposits would be described by the term "indicated." This term as defined applies to material for which estimates of the quality and quantity have been computed partly from samples and measurements and partly from reasonable geologic projections. The term "inferred" as defined by the Survey and the Bureau describes estimates of known but unexplored deposits for which estimates of the quality and size are based on geologic evidence and projection.

There is no intent here to replace the three P's as used by industry. As the basis for investment and planning decisions, such as the acquisition and sale of property, development of mining plans, etc, assessments in the framework of the three P's are useful and appropriate. However, they do not provide for the recognition and appraisal of unexplored but known deposits, much less for deposits that may be presumed to exist but are not yet discovered. It is from the unexplored and undiscovered deposits that much of our future supply will come, and from a national standpoint it is important that we have some evaluation of their magnitude.

Brobst and Pratt (1973) introduced two subdivisions of undiscovered deposits—"hypothetical" deposits that may reasonably be expected to exist in a known district, and "speculative" deposits that may occur in a broadly favorable geologic terrane where no discoveries have yet been made. Undiscovered petroleum accumulations can thus be hypothesized to exist in association with known geologic environments (the unexplored parts of a reef complex, for example), and we can speculate that oil and gas remain to be discovered in stratigraphic traps in incompletely explored but broadly favorable basins.

All the terms describing the degree of certainty about the existence of the deposits and the reliability of the estimates are also applicable to materials that are presently subeconomic. For example, during the exploration that led to the determination that a given deposit was too low grade to be producible, a considerable amount of material may have been measured. Also, many known deposits for which indicated or inferred estimates can be prepared are incompletely explored precisely because they have been judged to be subeconomic. Just as the spectrum of known deposits includes many that are too low in quality or too difficult to produce, undiscovered deposits can be expected to include some that are minable under present conditions and some that are not.

The "subeconomic" category is subdivided into paramarginal and submarginal classes to separate materials that now border on being economically producible from those that would require a substantially higher price or a major cost-reducing advance in technology to be economically producible.

The use of this classification will help delineate targets both for exploration and for technologic research and development, and will give some indication of our country's resource potential beyond the reserves that are presently being explored and developed. Except for those deposits in the measured category, the estimates necessarily have low reliability and should not be regarded as anything more than a dated assessment based on the state of existing knowledge. There are two important reasons for this. First, because the estimates depend on geologic projections, a great and unappraisable uncertainty is involved. A relatively small change in the thickness of a bed or the width or continuity of a vein can affect an estimate of the tonnage in a partly explored deposit by a factor of two or three. The second reason for the low reliability of estimates of potential resources is that we do not know enough about the earth's crust or about the frequency distribution of various kinds of mineral concentrations to know what may be geologically available.

Furthermore, we cannot forecast with much confidence the technologic advances or economic changes that may determine what is eventually producible. Today's measured reserves of nearly all minerals were once in some other category—and, at some time in the past, many of today's usable deposits would not have been reported in any category in this classification, simply because they were in a form not known to exist geologically or not imagined to be eventually usable commercially. For these reasons, I personally avoid using the term "inventory" in conjunction with resource appraisal. The term implies a fixed and appraisable stock of usable materials, when actually the stock is continually changing and what is presently known and imagined to be usable is assessible only to a very low degree of reliability.

Nevertheless, if the foregoing limitations and the continual addition of knowledge and of new developments are considered, an appraisal of national resources in the framework of this classification is valuable for many purposes. For purposes of appraisal of national resources, it would be useful to subdivide measured reserves into developed and undeveloped categories. Not only would such a breakdown be useful in indicating how much production could be accelerated, but it also would be an indication of the magnitude of the reserves that are fully explored and those that probably represent minimum estimates.

Although I believe the concepts of reserves and resources outlined here are valuable in understanding our reserve and resource position and potential, the difficult task is to develop the methods, procedures, and knowledge necessary to estimate reserves and resources beyond the proved or measured category. Fortunately, useful approaches have been developed, and this conference may do much to indicate the methods that will receive wide acceptance.

REFERENCE CITED

Brobst, D. A., and W. P. Pratt, eds., 1973, United States mineral resources: U.S. Geol. Survey Prof. Paper 820, 722 p.

U.S. Geological Survey Oil and Gas Resource Appraisal Group[1]

HARRY L. THOMSEN[2]

ABSTRACT   One of the primary roles of the U.S. Geological Survey is to provide mineral resource appraisals to assist in the determination of future government policy with respect to the leasing of public, Indian, and outer-continental-shelf lands. At various times in the past the Survey has made and published estimates of oil and gas resources in the United States and the world, but these assessments usually were carried out as short-term projects by individuals using different appraisal methods. Early in 1974, the start of a new project resulted in the formation of the Resource Appraisal Group. This group is charged with establishing a coordinated continuous system that will provide for the compilation and evaluation of all data pertinent to reliable resource assessment. The system being designed uses the geologic province as a basic unit. Primary emphasis will be on appraisal of the United States and its offshore areas, but studies also will be made of important provinces throughout the world. Procedures, analyses, and evaluations will be coordinated by a small group of professionals who will rely on the expertise of other geologists, geophysicists, and engineers to provide much of the basic information needed for ultimate evaluation of petroleum potential. The system in operation will draw heavily on the AAPG Oil and Gas Field Data Bank and Map Project, and plans are being made to use or build other computerized data banks to facilitate the updating and statistical analysis of province data.

INTRODUCTION

One of the responsibilities of the U.S. Geological Survey is to appraise mineral resources and to make the results of these appraisals available to government officials and the public. This responsibility has been met in the past for oil and gas through periodic publication of the results of special, short-term studies. Current needs, however, require new approaches to the problems of resource appraisal. In recognition of these needs, the Geological Survey in 1973 defined a new project which, early in 1974, gave rise to the Oil and Gas Resource Appraisal Group in the Branch of Oil and Gas Resources. This paper reviews the objectives, plans, and current status of that group.

OBJECTIVES AND PROGRAM

The general assignment of the Resource Appraisal Group (sometimes referred to as "the Group") is to develop better appraisal methods; to apply these methods to make estimates of the oil and gas resources of the various countries of the world—the United States in particular—on a long-range continuing basis; and to make these estimates available to interested government agencies and to the public. Specific objectives under this assignment include the following:

1. Complete a detailed appraisal of the most important provinces within 3 years, using the geologic province as the basic unit for analysis.

2. Within the first year, make a preliminary appraisal of all important petroleum provinces in the world. This appraisal would serve as a framework for assigning priorities for detailed province evaluations.

[1]Manuscript received, February 3, 1975.
[2]U.S. Geological Survey, Denver, Colorado 80225.

3. Develop and utilize a comprehensive information system that will facilitate collecting, storing, retrieving, sorting, listing, characterizing, and correlating basic data. This system will consist not only of computerized data, but also of province folios, maps, reports, and excerpts of published articles.

4. Devise a method of allocating appraisals to public and Indian lands.

5. Provide a system of reporting to keep interested people up to date on progress. Terminology used is to conform to the mineral-resource terminology agreed upon by the Bureau of Mines and the Geological Survey. A range of values will be used to convey the uncertainty of resource appraisal figures.

6. Build a staff of 12-15 professionals to form the permanent core of the Group. It is our intention to design the program so that discrete "packages" of critical information can be assembled for us from sources outside the Group. The primary functions of the Group will be to summarize and analyze this information, make the final resource appraisals, update and revise data as necessary, and report results.

CURRENT SITUATION

We have made a good start on the program outlined above. The preliminary appraisal of important petroleum provinces throughout the world is under way and essentially on schedule. Most of our effort is being directed toward an evaluation of the oil and gas potential of the United States and its outer-continental-shelf areas, but a cursory definition and an appraisal of provinces in the rest of the world are being carried out concurrently. We plan to use volumetric-geologic methods to make the initial appraisals.

The Resource Appraisal Group is the beneficiary of a substantial computerized oil- and gas-pool data bank which is being compiled in Norman, Oklahoma, through the joint efforts of the AAPG and the USGS. Also, we are in the process of purchasing Petroleum Informations's computerized well-history file and are hopeful of obtaining the use of other important banks of computerized data through gifts, purchase, or trade. As a result of preliminary discussions, we also are confident that we can obtain access to important collections of basic geologic information on many of the important provinces throughout the world. One of our biggest unfilled needs is for reserve and production data listed by province and by stratigraphic unit.

We have attempted to solve the problem of allocating appraised resources to onshore public and Indian lands in the United States by designating the county as a basic subdivision of the province.

We have not decided on a standard procedure for keeping interested people abreast of our progress, but are confident that this problem will essentially solve itself as our efforts begin to produce meaningful results.

When the Resource Appraisal Group was organized in February 1974, it consisted of four professionals and a secretary. By August, the staff consisted of eight geologists (three of whom are part-time), one computer programmer, a librarian, and a secretary, as well as three students hired for the summer. Hopefully, two or three professional people will be added soon, and we expect to amplify the effectiveness of the Group by reaching out to various people within the Survey to help with temporary assignments for which they are particularly well qualified. Also, the Group is looking for senior geologists with extensive experience in

specific geologic areas who would be interested in temporary employment
with the Survey for the purpose of assembling the basic information
needed for province evaluations.

DESIGN OF A PETROLEUM-PROVINCE ANALOG SYSTEM

Members of the Resource Appraisal Group are in general agreement
that long-range efforts should be directed toward building a comprehensive petroleum-province analog system (Fig. 1) based on a systematic
collection and evaluation of data from hydrocarbon-bearing provinces
throughout the world.

The first phase in building this system consists of collecting basic
geologic data for each province, interpreting and characterizing these
data, and summarizing them on evaluation forms. As part of this procedure, we are subscribing to several established information services.
The second and third phases provide for similar treatment of data from
field studies and from basic information on production, reserves, and
character of fluids. Much of the work in these three phases can be

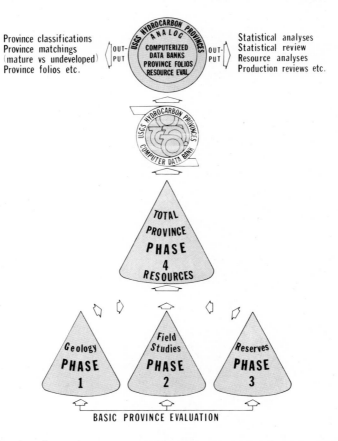

FIG. 1—Overview of a petroleum-province analog system based on a systematic collection and evaluation of data from hydrocarbon-bearing provinces throughout the world. (Diagram designed by Betty M.
Miller, Resource Appraisal Group.)

handled by outside professionals following specifications set by the permanent Resource Appraisal Group. The fourth phase involves a total hydrocarbon-province evaluation based on information from the first three phases. This work would be handled by members of the Group and would include, as a major element, an appraisal of the resources of each province by the best available method.

The fifth and final phase would consist of building computerized data banks with data from the first four phases, and using these banks—together with general information from province reports, folios, and resource appraisals—to build better analogs.

Admittedly, this is an ambitious and idealized plan but, if this system is followed, the framework of information constructed will improve with time and should provide a solid foundation upon which to build resource appraisals.

Estimating Resources of Crude Oil and Natural Gas in Inadequately Explored Areas[1]

THOMAS A. HENDRICKS[2]

ABSTRACT   Estimating resources of crude oil and natural gas contained in inadequately explored regions is fraught with difficulty. An analog method that I have used is described as follows.

1. Compute the total oil and gas resources originally in place in the intensively explored parts of the United States on the basis of (a) past production, (b) proved reserves, (c) future revisions of proved reserves, (d) future extensions of known pools, (e) additions in pools to be discovered in the course of development drilling, and (f) subcommercial accumulations or penetrated but undetected deposits that will be brought into production in the future.
2. Divide this explored area into a graded series of categories of total potential in barrels per 1,000 sq mi (2,590 km^2).
3. Classify the various parts of the inadequately explored areas of the world into similar categories on the basis of past production and proved reserves (if any) and geologic characteristics.
4. Measure the area in each category in the inadequately explored region and multiply by the factor established for the geologically similar explored area.

   Much of this procedure is based on available factual data, but the ratings of both the explored and the inadequately explored areas based on geologic factors are, to a large degree, subjective, and so is the decision as to what constitutes adequate exploration. To permit more complete judgment of the procedure, a map of the sedimentary basins of the world currently is being prepared for publication.

INTRODUCTION

Back in the Dark Ages, a man, whose name was not recorded in history, spent a year dissecting the carcass of an elephant. He did not do it very well and contributed nothing to the knowledge of anatomy. However, he achieved a measure of local fame for being willing to tackle a big and odorous job. In attempting to set up a method of estimating the resources of liquid hydrocarbons and natural gas in underdeveloped areas, the writer feels a certain kinship to that unsung individual. Such an effort is hazardous—if it were not, there would be no need for this conference.

The problems of estimating oil and gas potential differ markedly from those of estimating other fossil fuels. Coal, oil shale, and asphalt occur in solid bodies that are extensively explored at the earth's surface, where their dimensions, character, and geologic relations may be observed directly. Oil and gas accumulations, by contrast, are deeply buried; moreover, they occur as fluids filling fractures and intergranular spaces in the rock through a much wider range of stratigraphic settings than do the solid fuels. In short, resources of the solid fossil fuels are at least partly visible and relatively concentrated, whereas resources of oil and gas are totally unseen and somewhat diffuse.

Because man cannot see oil and gas resources, it is not surprising that he has been ultraconservative in estimating their undiscovered quantities. Repeatedly, expert estimates of total quantities that are undiscovered have been exceeded within a decade or so by quantities actually discovered. This conservatism regarding the unseen is also reflected in estimates of the amount of unproduced hydrocarbons in accumulations

[1] Manuscript received, January 6, 1975.
[2] U.S. Geological Survey, Denver, Colorado 80225.

that have already been outlined by drilling (i.e., estimates of proved reserves). The record shows that estimates of ultimate recovery from drilled pools are generally revised sharply upward as cumulative production mounts with time.

Geologic science has been very successful in classifying portions of the earth's crust as favorable, unfavorable, or virtually impossible for the occurrence of oil and gas, but the exact location and extent of the occurrences can be determined only by drilling. Furthermore, only extensive drilling and production experience can provide the statistical sample needed to attempt quantitative estimates of the producible resources in the favorable unexplored parts of the earth's crust. Moreover, much of the purely geologic information needed for intelligent appraisal of unexplored rock can be provided only by drilling.

At present, only a tiny portion of the geologically favorable rock in the earth's crust has been explored by drilling. The amount of exploratory footage drilled in the United States dwarfs the total in the rest of the world, but even in the United States there is "room" in the geologically favorable rocks for much more exploratory drilling. Until a far larger "sample" has been drilled, estimates of undiscovered quantities will be subject to major revision periodically.

GEOLOGIC–ANALOG METHOD

A method for making such estimates, devised in 1964 and 1965 by A.D. Zapp, J.F. Pepper, and T.A. Hendricks (Hendricks, 1965), may be described as a geologic-analog method. It consists of six main steps:

1. Estimating the amount of oil originally in place in a relatively well-explored region that has considerable geologic diversity. The lower 48 states and the continental shelf off southern California and in the Gulf of Mexico were chosen for this step. One particular asset of this selection at that time was the availability of a published study by the Interstate Oil Compact Commission on oil in place in known reservoirs in the United States. This function now has been taken over by the American Petroleum Institute (API) and a report is published annually.

2. Classifying various parts of this large area into categories on the basis of the amount of liquid hydrocarbons and natural gas present in each in terms of a factor of millions of barrels per 1,000 sq mi (2,590 km^2). The classification is based on the assumption that past production and proved reserves of crude oil are proportional to total oil in place. A related factor was derived for natural gas and natural gas liquids. It is realized that some basins will contain only oil, whereas others may contain only gas, either dry or associated, and that some apparently favorable basins will be barren.

3. Classifying the underexplored areas of the world into the same categories as those of the United States on the basis of geologic similarity and known hydrocarbon occurrences.

4. Measuring the area of each underexplored region.

5. Multiplying the area in units of 1,000 sq mi (2,590 km^2) by the range of factors developed for basins in the same category in the United States.

6. Adding the totals for areas in each category for the country or sedimentary basin for which the information is needed.

Although this method sounds simple, it is fraught with uncertainty. It is apparent that a single numerical value for a basin could not be expected to be accurate. Therefore, each of the values used for each commodity should be considered a median in a range from 50 to 150 percent of each number used. This part of the procedure was not followed in U.S. Geological Survey Circular 522 (Hendricks, 1965).

Step one of this procedure consists of at least eight components, which are themselves subject to some assumptions and are only partly supportable estimates. These components are (1) past production; (2) proved reserves in known reservoirs; (3) future revisions due to increased percentage recovery; (4) future revisions resulting from net increases in estimates of oil in place in drilled acreage; (5) future extensions to known pools through development drilling; (6) new pools to be discovered in the course of development drilling of known pools; (7) new pools in originally subcommercial deposits and in drilled but undetected deposits that will be brought into production in the future; (8) future discoveries by exploratory drilling.

Even past production is not a measured quantity. Allowance must be made for blowouts, for flaring and venting of natural gas and associated liquids, and for gas used for pressure maintenance and lease operations.

Future revisions of estimates, extensions of known pools, and discoveries during development can be estimated with considerable accuracy by using the historical record. API determinations of oil added to proved reserves by revisions and extensions are now credited back to the year of discovery of the pools involved.

Recompletions that permit production of initially subcommercial or undetected accumulations are more difficult to determine, but it can be assumed with some assurance that any important additions for this oil will be included in revisions and extensions.

Future discoveries and the amount of hydrocarbons that will remain undiscovered are dependent on the amount of drilling that can be done economically. Assumptions must be made as to the amount of drilling that would be required to discover all, or nearly all, important deposits. The volume of "favorable" rocks in the United States that have been adequately explored and the volume of rocks remaining to be explored are relatively well known. From these data, reasonable assumptions can be made as to the ultimate amount of exploratory drilling that can be done at a profit and the number of discoveries to be expected. The general decrease in the amount of oil and gas discovered by each 1,000 ft (305 m) of exploratory drilling in the last 15 years suggests that the incidence of crude oil in the unexplored rocks ranges from the average in the explored rocks for the richest fraction to nil for the poorest fraction.

The classification of little-explored regions into categories equal to the better-explored counterparts in the United States is particularly difficult. Some part of this procedure must be subjective, reflecting the experience and prejudices of the persons making the evaluation. Some of the factors that are logically considered are (1) abundance of hydrocarbons; (2) composition of hydrocarbons; (3) recognition of whether source beds are marine or nonmarine; (4) recognition and evaluation of reservoir beds; (5) sedimentary history of the area—deltas, evaporite basins, reefs, isopach thins, unconformities; and (6) structural history of the area.

This list could be expanded manyfold and each item could be assigned a weighted value, but the evaluation would still be limited by the competence and background of the evaluators. For this reason, the best results should be expected from studies by teams made up of competent individuals with varied backgrounds.

Many discoveries that would be commercial in the United States are noncommercial elsewhere at present because of remoteness, lack of facilities for production and transportation, and location in a hostile environment where costs are high. However, discoveries that would be uneconomic elsewhere may be very profitable if made in heavily industrialized areas such as Japan or Western Europe. Adjustment for these factors must be made in arriving at the proper category to be assigned to individual basins.

In areas where exploration has been extensive and production has been prolific, classification by analogy is unnecessary. For example, the data on past production and proved reserves in the Persian Gulf regions, Venezuela, and Western Canada are far more significant than appraisal by analogy. Therefore, appraisal by statistical analysis is more effective. Other areas characterized by multiple giant fields—such as the North Sea, Algeria, and Nigeria—may attain this status.

CONCLUSIONS

The difficulties inherent in appraising the potential of any one undeveloped basin or part of a basin are so great that the probability that the basin will produce hydrocarbons *within the range* assigned to it by analog classification is probably no more than 50 percent. Indeed, some basins that appear favorable geologically may yield no hydrocarbons. However, if the region under consideration is large and diverse enough, compensating factors should cause the average for that large area to fall within the range of the potential determined by analog comparison.

For these reasons, the U.S. Geological Survey plans to prepare world-wide maps of all basins, both on land and offshore, together with an appraisal of each. However, it is clearly understood that only the average potential of major geologic provinces can be expected to be significant.

REFERENCE CITED

Hendricks, T. A., 1965, Resources of oil, gas, and natural gas liquids in United States and the world: U.S. Geol. Survey Circ. 522, 20 p.

Accelerated National Oil and Gas Resource Appraisal (ANOGRE) [1]

WILLIAM W. MALLORY[2]

ABSTRACT   The purpose of the ANOGRE system is to derive estimates of recoverable and subeconomic undiscovered oil and gas resources in the lower 48 states. Unproduced recoverable hydrocarbons can be classified, by decreasing degree of confidence, into (1) measured (proved) reserves, (2) inferred reserves, and (3) undiscovered recoverable resources.

The bulk of unproduced hydrocarbons can be classified as the sum of measured reserves, inferred (and indicated) reserves, and undiscovered recoverable resources. Inferred reserves are equivalent to about 50 percent of measured reserves. The quantity of known recoverable hydrocarbons is to the volume of drilled rocks (both dry and productive) as the quantity of undiscovered recoverable resources is to the volume of undrilled favorable (potential) rocks times a numerical richness factor f. The quantity of known recoverable hydrocarbons is defined as the sum of cumulative production plus measured and inferred reserves.

Determination of the volume of drilled and potential rocks involves areal and vertical classifications. The volume of potential rocks is defined as the volume of favorable rocks minus the volume of rock (1) occupied by the stratigraphic unit containing an oil or gas pool, (2) attributable to extensions, revisions, and higher and lower "pays" not yet defined, and (3) proved barren by dry holes.

The U.S. Geological Survey's Oil and Gas Pool Data Bank, presently the most comprehensive source of digitized United States pool data, is used extensively in making these computations.

ANOGRE SYSTEM

The ANOGRE system was developed as a documented procedure by which estimates of recoverable and subeconomic undiscovered oil and natural gas resources for the onshore conterminous United States can be derived and continuously updated using computerized data techniques.

The specific rationale of the ANOGRE system can be stated as follows. Unproduced recoverable hydrocarbons can be classified into three categories by decreasing degree of confidence in their existence: (1) measured (proved) reserves, (2) inferred reserves, and (3) undiscovered recoverable resources. *Measured reserves* are published yearly by the American Petroleum Institute (API) and the American Gas Association (AGA). Analysis of past records suggests that *inferred reserves* for a given year for the entire United States are equivalent to about 50 percent of measured reserves, but other more precise methods of determining this quantity for individual geologic provinces are being developed and may be substituted. *Undiscovered recoverable resources* (those present in undrilled wildcat areas) are computed on the basis of reasoning that the ratio of the quantity of undiscovered recoverable resources to the volume of undrilled favorable rocks is proportional to the ratio of the quantity of known recoverable hydrocarbons to the volume of drilled rocks times a factor f. These assertions and their computational developments for all the regions prospective for oil and gas in the lower 48 states are stated in equations 1 to 4.

[1] Manuscript received, March 10, 1975.
[2] U.S. Geological Survey, Denver, Colorado 80225.

Unproduced Hydrocarbons

$$HC_{\text{unproduced}} = HC_{\text{measured reserves}} + HC_{\text{inferred reserves}} + HC_{\text{undiscovered recoverable resources}} \quad (1)$$

Inferred Reserves

$$HC_{\text{inferred reserves}} = HC_{\text{measured reserves}}\,(0.50\pm) \quad (2)$$

Undiscovered Resources

$$\frac{HC_{\text{known recoverable}}}{V_{\text{drilled}}} = \frac{HC_{\text{undiscovered recoverable}}}{V_{\text{potential}}}\,f, \quad (3)$$

or

$$HC_{\text{undiscovered recoverable}} = \frac{HC_{\text{known recoverable}}}{V_{\text{drilled}}}\,(V_{\text{potential}})\,f, \quad (4)$$

where HC = quantity of hydrocarbons and V = volume of rock.

Equation 1 indicates that the bulk of unproduced hydrocarbons can be classified as the sum of measured reserves, inferred (and indicated) reserves, and undiscovered recoverable resources. Equation 2 indicates that inferred reserves are equivalent to about 50 percent of measured reserves. Equation 3 states that the quantity of known recoverable hydrocarbons is to the volume of drilled rocks (both dry and productive) as the quantity of undiscovered recoverable resources is to the volume of undrilled favorable (potential) rocks times a numerical richness factor f. Equation 4 is a restatement of equation 3.

The crux of the problem at this point, then, is the determination of the quantities of known recoverable hydrocarbons, volume of rock drilled (both productive and barren), volume of *potential rock*, and f. *Potential rock* is defined as rock lithologically and tectonically favorable for hydrocarbon production, but as yet undrilled. Determination of the numerical value of f is under study. At present the approach to this factor is based on the quantity of hydrocarbons found per foot of exploratory drilling as a function of cumulative exploratory footage.

The *quantity of known recoverable hydrocarbons* is here defined as the sum of cumulative production plus measured and inferred reserves. Cumulative production is derived from the U.S. Geological Survey's computerized Oil and Gas Pool Data Bank, and measured reserves are taken from the API/AGA reserves volume. For logistic reasons, the county has been selected as the basic geographic statistical unit. Because measured reserves are reported only by states, they are allocated to counties on the computational basis that there is a simple linear relation between county annual production (retrieved from the Data Bank) and state reserves. The relation can be expressed thus:

$$\text{County reserves} = \frac{\text{County annual production}}{\text{State annual production}}\,(\text{State proved reserves}).$$

Determining the volume of drilled and potential rocks is a more complex process. Figure 1, a map of the United States, shows in heavy lines geologic provinces as defined by The American Association of Petroleum Geologists Committee on Statistics of Drilling (CSD). The province boundaries have been adjusted to county lines to facilitate digital manipulation. Regions which do not produce oil and gas in commercial

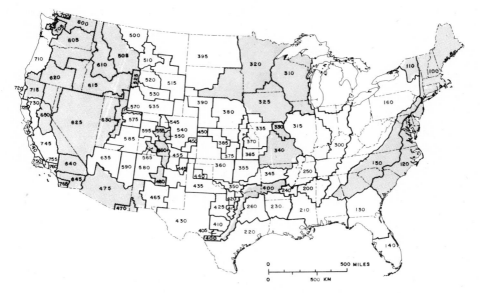

FIG. 1—United States, showing geologic provinces as mapped by the AAPG
Committee on Statistics of Drilling.

quantities (with minor exceptions), and which are not likely to do so
in the future, are shaded.

Regions shown in white on Figure 1 contain the oil and gas fields of
the onshore 48 states. All future onshore discoveries (with possible
minor exceptions as, for example, in the Great Salt Lake basin) are expected to be made in these prospective regions.

Second-order areal classification of rocks in these regions is by
geologic provinces, which are named and number-coded (Meyer, 1968).
There is a total of 60 provinces in the prospective region. Third-order
areal classification is by county, a unit small enough to be useful
for statistical manipulation. Because oil and gas data are commonly
reported by state, CSD province, Petroleum Administration for Defense
(PAD) district, or other political unit, a county breakdown makes it
possible to deliver data computations in any needed geographic combination.

Within the vast central cratonic region and the Gulf coastal plain,
a vertical classification of the sedimentary rocks is made by age. From
the Pleistocene to the base of the Mississippian System, classification
units coincide with the series. In strata below the Mississippian, the
rocks have been grouped for pragmatic reasons as follows: Upper Devonian
Series, Lower and Middle Devonian Series (combined), Silurian System,
Middle and Upper Ordovician Series (combined), and Cambrian System and
Lower Ordovician Series (combined as Cambro-Ordovician).

For convenient reference, each rock layer is here called a *major
stratigraphic unit* (MSU). Thirty MSU's are recognized. The lithofacies
and thickness of the rocks in each unit in the central cratonic region
and the Gulf Coast provinces are shown on 30 maps at a scale of 1:5,000,000
(on file in Denver). The collection of maps is basic to a project of
this kind and constitutes a major part of the system. Only two relatively
minor exceptions to this mapping plan are of necessity incorporated. The
exceptions concern the tectonically complex West Coast basins and the

stratigraphically obscure Cenozoic rocks in the intermontane Rocky Mountain basins.

The principal sources for compiling the map folio are (1) the paleotectonic map of the U.S. Geological Survey, (2) the *Geologic Atlas of the Rocky Mountain Region,* published by the Rocky Mountain Association of Geologists (Mallory, 1972), and (3) a collection of maps compiled from the literature for this study by Edward Sable, Katharine Varnes, Wallace DeWitt, and others.

In all prospective regions, favorable rocks in each MSU were mapped as follows. A transparent matte-surface cronaflex positive showing the CSD province outlines was placed over each of the 30 thickness and lithofacies maps. Geologists familiar with specific regions critically evaluated the rocks in each MSU for known and possible undiscovered hydrocarbon content and compared them concomitantly with the oil- and gas-field and surface geologic maps of the United States. Favorable rock areas then were outlined in colored pencil on the overlay maps. In addition, the *Tectonic Map of North America* (King, 1969) and the *Tectonic Map of the United States* (Cohee et al, 1969) served as major regional guides to tectonics and diastrophic history. The CSD-province overlay was applied also to the surface geologic map of the United States and to the oil- and gas-field map to maintain surface and subsurface provincial orientation on all maps at all times. Judicious use of the stratigraphic maps makes it possible to dilineate, MSU by MSU, the *favorable rocks* of the lower 48 states. *Favorable rocks* are here defined as rocks which contain oil and gas in known fields and rocks which, because of their lithology, tectonic history, and general location, reasonably can be expected to contain undiscovered oil and gas.

Criteria for ascertaining what rocks are favorable include:

1. The presence of oil and gas fields in the mapped stratigraphic unit.
2. A cratonic or coastal-plain setting.
3. Depth less than 25,000 ft (7,619 m) but greater than 500 ft (152 m).
4. The presence of marine rocks with considerable volumes of carbonate rock and sandstone.
5. The presence of special rock associations such as reefs, evaporite complexes, dark shales, critical pinchouts, and megafacies changes.
6. A suitable tectonic environment.
7. Favorable structural belts containing such features as growth faults, salt domes, and ridges.

Negative indications such as massive redbeds and conglomerates, freshwater saturation, excessive depth of burial, shallow burial, and continental lithology were also carefully noted. Regional geologic background, tempered with judgment, and experience in the geology of oil and gas governed the actual delineation of favorable areas. The result is a breakout of rocks favorable to the presence of oil and gas pools.

After favorable rocks had been mapped for all MSU's in the country (with the exceptions mentioned), the thickness of each stratigraphic unit in every county in the prospective regions was tabulated. These figures were then keypunched and stored on discs to become the ANOGRE Rock Census, a versatile data subset.

The *volume of potential rocks* is defined as the volume of favorable rocks minus the following volumes:

1. The volume of rock occupied by the stratigraphic unit containing an oil or gas pool (V_{pool});

2. The volume of rock attributable to extensions, revisions, and higher and lower "pays" not yet defined ($V_{inferred}$); and
3. The volume of rock proved barren by dry holes (V_{dry}).

These three rock volumes, must, therefore, be computed for each county and MSU in the prospective regions.

Figure 2 is a block diagram of a hypothetical county whose boundaries have been extended downward to basement (or the economic floor). The reservoir is black and V_{pool} is ruled. Although only one pool is shown, the vertical-ruled area represents diagrammatically the sum of all pool volumes in the county. This rock volume is computed from the Survey's Oil and Gas Pool Data Bank. Figure 3 shows by ruling the volume of rock attributable to extensions, revisions, and higher and lower "pays"; this volume of rock contains inferred reserves and hence is called *inferred rock*. (The pool cylinder includes all MSU's contained in the Rock Census, which contains only favorable rocks.) Figure 4 emphasizes a cylinder of rock surrounding a dry hole. Studies by the Geological Survey indicate that it is statistically logical to consider that a dry hole condemns an area about equal to the size of the average field in a given geologic province.

Because $V_{potential}$ is equal to $V_{favorable}$ minus V_{pool}, $V_{inferred}$, and V_{dry}, the ruled part of Figure 5 indicates the volume of $V_{potential}$. All of these quantities are available in digitized form and are derived from the following sources:

V_{pool}:   from the Oil and Gas Pool Data Bank

$V_{inferred}$: from the Data Bank and the Rock Census

V_{dry}:   from dry-hole information, purchased in digital form from a commercial source, combined with data from the Rock Census.

$V_{favorable}$: from the Rock Census.

The basic equation as restated (equation 4) now can be solved by substituting the quantities described for the generalized items, to give:

$$HC_{ur} = \frac{HC_{cum} + HC_{mres} + HC_{inf}}{V_{pool} + V_{inf} + V_{dry}} \; V_{fav} - (V_{pool} + V_{inf} + V_{dry}) f \; ,$$

where

HC_{ur} = undiscovered recoverable hydrocarbons,

HC_{cum} = cumulative production of hydrocarbons,

HC_{mres} = measured reserves of hydrocarbons,

HC_{inf} = inferred reserves of hydrocarbons,

V_{pool} = volume of pool rock,

V_{inf} = volume of inferred rock,

V_{dry} = volume of dry rock,

V_{fav} = volume of favorable rock,

f = the richness factor.

Solution of the equation yields undiscovered recoverable oil by MSU per county. Summations can be made for any geographic combination—for example, the undiscovered recoverable oil or gas attributable to Mesozoic rocks in a state or a geologic province, or in all of the lower 48 states.

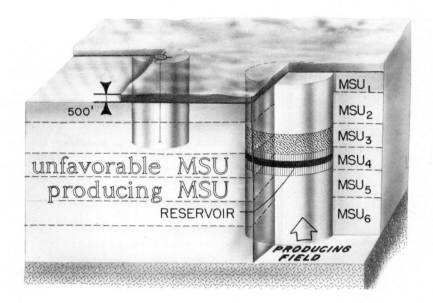

FIG. 2—Volume of pool rock (vertical ruling).

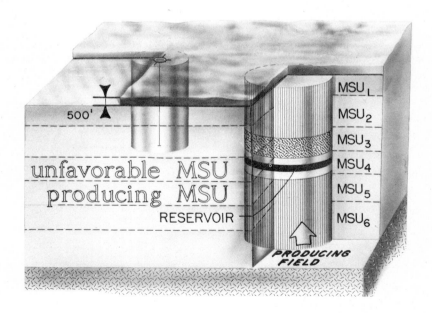

FIG. 3—Volume of inferred rock (vertical ruling).

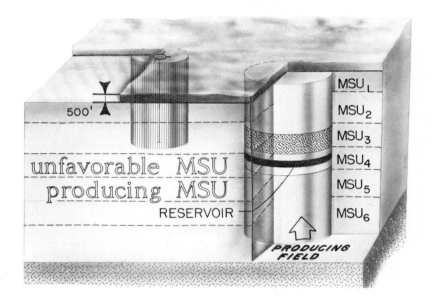

FIG. 4—Volume of dry rock (vertical ruling).

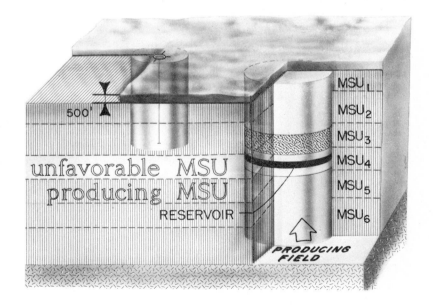

FIG. 5—Volume of potential rock (vertical ruling).

Subeconomic resources can be derived by dividing the recoverable figures by the appropriate recovery factor; original oil in place is derived by finding the sum of cumulative production, flared and vented fluids, measured reserves, inferred reserves, undiscovered recoverable resources, and subeconomic resources.

In order to render these computations significant, exhaustive effort and considerable funds have been expended to provide the most complete file of rock and hydrocarbon data available. The Oil and Gas Pool Data Bank is at present the most comprehensive source of digitized United States pool data in existence available to government. Rock-volume figures are based on the map folio described. Dry-hole data are purchased.

In a Department of the Interior news release dated March 26, 1974, the Survey presented resource figures using f ranging from 1.0 to 0.5. The resulting figures are higher than those generally supported by industry. New attention to the value of f and its mode of application to the basic equation are under study by R. F. Mast, T. H. McCulloh, and L. J. Drew at this writing, and preliminary results indicate that lower values for f may be justified. New studies of undiscovered resources currently under way will utilize these revised values.

REFERENCES CITED

Cohee, G. V., et al, 1961, Tectonic map of the United States: U.S. Geol. Survey and AAPG, 2 sheets, scale 1:2,500,000.

King, P. B., 1969, Tectonic map of North America U.S. Geol. Survey, 2 sheets, scale 1:5,000,000.

Mallory, W. W., ed., 1972, Geologic atlas of the Rocky Mountain region: Rocky Mtn. Assoc. Geologists, 331 p.

Meyer, R. F., ed., 1968, Geologic provinces map: AAPG, revised 1974.

Potential Petroleum Resources—Classification, Estimation, and Status [1]

LEWIS G. WEEKS[2]

ABSTRACT   Reasonably reliable resource estimation is basic to planning and investment decisions in petroleum exploration. The basic need is to improve this reliability, but it is doubtful that this can be accomplished in a constantly changing resource situation by introducing an extensive array of vaguely defined, questionably contributory, and time-consuming subcategory terminologies.

Similarly, the use of artificially chosen yardsticks has no assured value as the prime basis for estimating potential resources, although such yardsticks are useful for quantitatively representing or comparing the volumes or incidences of estimated occurrence. Geologic criteria, including geologic interpretation, are the only criteria having any measurement value in regard to the results of the worldwide exploration experience of the industry. Many factors determine or affect the incidence of petroleum occurrence.

Worldwide interest in world resources and their estimation, particularly that of energy, dates back to the founding of the United Nations. A conservative projection of the upward trend of overall world energy demand in the past 20 years indicates what the total demand might be during the next 100 years, if supplies of energy and all of the other required industrial raw materials are, or could continuously be made, readily and economically available. Projections extending so far into the future involve many elements of uncertainty other than those of availability; not the least of these are economic and political variables. Nevertheless, any reasonably possible projection of potential petroleum supply from conventional sources shows that they may reach their peak of production near or soon after the turn of the century; at that time petroleum will provide about 10-15 percent of the indicated potential world energy demand. It will supply a similar proportion of United States energy demand, and the greater part of it will be imported.

By the year 2020, the amount of energy supplied by petroleum will have decreased to about 6-8 percent of the larger world demand. Conventional petroleum supply will continue to decline for perhaps another 100 years, ending in a comparative trickle, although it still will be in demand. Its contribution to the indicated total energy demand over a period of 100 years from the present will be less than 4 percent.

Petroleum is unique as an energy and raw-material source by reason of its vast and unparalleled utility and convenience. For all forms of transportation, it is the only practicable energy source. It is unsurpassed not only as an energy source, but also as a raw material for a great host of industry uses because of its chemistry and end-product versatility. Above all, the cost of petroleum is still, and probably will continue to be, unmatched by any of the other presently known sources.

CLASSIFICATION TERMS

Three very appropriate, meaningful terms have been used to classify our petroleum resources; they are "proved reserves," "potential resources," and "oil in place." Conventional proved reserves and potential resources, plus any secondary recovery, are normally derived from the amount of oil in place. These terms are simple and broadly understandable by everyone, including the layman. Any subcategories should be kept to a minimum. Introduction of a list of vaguely defined category subdivisions and fancy new names will not enable us to improve the accuracy of our estimates; and that is our basic need.

We are now in an age when the use of fancy new terms, or of old terms given new meanings, is in vogue—for example, "reserve life indices" instead of "years' supply ratio," "scenario" instead of "schedule," "basin typing and modeling" instead of "basin classification," and "net recoverable reserves" instead of "proved reserves." Other proposed classes of resources are "inferred," "hypothetical," "speculative," "paramarginal,"

[1]Manuscript received, January 28, 1975.
[2]Lewis G. Weeks Associates, Westport, Connecticut 06880.

etc, but definitions vary so that it is impossible always to ascertain
the meanings.

The only legitimate guidelines to resource understanding are (1) the
geologic knowledge about the particular resources, as viewed against the
background of world geology and exploration experience, and (2) the economic feasibility of recovery. Though these qualifications are simple
and widely comprehensible, even they are subject to change.

The meaning of the term *proved reserves* is clear to all. It pertains
to the petroleum (oil or gas) which has been found and which can be produced and utilized at a profit. It corresponds to what the shopkeeper
has available on his shelves for profitable sale and use.

Potential petroleum resources comprise the accumulations of oil and/or
gas which the geology alone indicates may be found in the future and converted into reserves that can be produced and utilized profitably at that
time. As with all raw materials, petroleum cannot be considered a reserve
or a potential resource unless it can be recovered and used economically.
Nonambiguous estimates of potential resources that are soundly based in
geology and experience have a special place in industry, but they need
to be reviewed as new facts appear.

QUANTITATIVE CLASSIFICATION OF BASIN AREAS

One criticism of resource estimates is that they should not be given
in approximate figures but as a maximum-minimum range of possibilities.
I believe that a single approximation by an experienced geologist is far
more reliable than a wide range of figures that may be based on questionable, lazily derived guesses. An estimate that is given, for instance,
in a range from zero to as much as several billion barrels does not speak
highly of the geologist's understanding either of the geology or of oil
occurrence.

Reasonably reliable resource estimation is basic to planning and
investment decisions. Our need is to improve this reliability, and I
am not ready to admit that we are unqualified to do it.

Although we are not beholden to a fickle public or to its vote-seek-
ing representatives, the public is interested mainly in the overall supply,
and we should try to present our estimates in understandable terms. As
with all estimates, however, those of our potential resources need to be
revised if and when new information makes that possible. Errors tend to
cancel out on a worldwide, or even countrywide, basis.

The main purpose of quantitative classification of basin areas, in
terms of the amount of petroleum it is thought each area will ultimately
produce, is to have the most definitive measure possible for investment
decisions. Resource estimations are aimed at efficiency in spending the
exploration budget. Their use in selecting the area or areas in which
the budget can best be spent makes such assessments much more important
than estimates of proved reserves.

The geologic research and comparisons that are necessary in making
reasonable estimates for many scores of basins or basin areas not only
force thinking and analysis, but they give a tangible meaning to the
rating of areas, basins, or regions. They give in quantitative terms a
very definite basis for discussion. Such a basis is not provided by the
obviously unstudied, so-called "estimates" commonly seen, nor is a real
basis provided by such widely interpretable, noncommittal, nonquantitative
forms of rating as "good," "fair," or "A, B, C." Overly optimistic state-
ments concerning potential resources, based at times on wishful thinking,
can be as misleading as, or more misleading than, some of the ill-advised
understatements that have been made.

Much has been learned about earth and basin history and the related facts of oil occurrence that commonly are shown in perspective to have been considered inadequately. On the other hand, we are apt not to appreciate that (1) high and low estimates of potential resources tend to cancel out on a worldwide basis, and (2) the extent to which total world potential estimates may prove to be too high or too low is unlikely to make any important difference in the proportion of potential world energy demand that petroleum will ultimately be able to supply. Too much of our attention may be monopolized by the review of world estimates or those of major countries. The one critical consideration is what we can develop to take over—adequately, economically, and efficiently—the load that has been so well borne by petroleum, which already is in short supply.

To compare or rate basins, basin areas, or countries *primarily for investment purposes*, I commonly have used yardsticks such as barrels of oil per square mile, per cubic mile, per discovery well, or per unit of investment. I continue to see references to these yardsticks (called the "Weeks method of estimating potential oil resources"), as well as the actual use of such yardsticks as the primary basis for estimating petroleum resources. Although my estimates, whether for the world or for small subdivisions, were made after careful study of each basin's geology interpreted in the light of the worldwide exploration experience of industry, all of my earliest estimates were incomplete and very conservative. This fact has been disregarded by those interested in making comparisons of individual estimates without regard for the purposes behind those estimates. These estimates of mine were *not* made for publication, but for the purpose of quantitatively rating basin areas for company investment decisions or, during the war, for government use.

RANGE OF OIL OCCURRENCE IN SPACE AND TIME: USE OF YARDSTICKS

One way in which potential resource estimates have been made is by projecting artificially chosen figures of average oil production per cubic mile (cubem) or cubic kilometer (cubek). The fallibility of this method may be shown in many ways. Potential productivities of basins range from less than 1,000 bbl to several million barrels of oil per cubem of basin sediments. Primarily, the use of barrels of oil per cubem or per cubek provides a means of comparison or rating of basins. Estimate comparisons of scores of basins or areas worldwide are a powerful means of checking on one's thinking.

The percentage of a basin that will prove commercially productive may range from less than 0.1 percent to as much as 5 or even 10 percent, and yields per proved acre may range from as little as a few hundred barrels of oil to as much as a million barrels, as in certain California and Middle East fields. So, while I concede that some yardstick method is useful for *comparing* or *rating* basins, it may be very misleading to apply any of these methods indiscriminately to the direct estimation of the production potential either of individual areas or of large basins or countries. A basin of 10,000 sq mi (25,900 km^2) area having a favorable environmental history may produce vastly more oil than one 10 times larger without such favorable history. Prolific oil basins exist in which 90 percent of the commercial oil has accumulated in a small fraction of the basin. Some of the very excessive estimates of undiscovered oil resources in the United States and elsewhere have resulted from applying to the entire area of the basins the same incidence of occurrence that is present in the areas and trends in which most of the petroleum is found. Many apparently third-rate structures, properly positioned in a favorable basin, have pooled many times more oil than have textbook-

perfect structures that are poorly situated in space or time, even in
the same basin.

Gas occurrence is equally variable, and the ratio of gas to oil occurrence varies fully as much as does its incidence. It varies from basin
to basin and from country to country, from a few hundred to tens or even
hundreds of thousands of cubic feet per barrel of oil (Weeks, 1962).
Clearly, it varies with the environment of deposition, the provenance
and nature of both the organic and the inorganic sediment constituents,
and the environmental history.

Some geologists have used the results of exploration in the United
States as an average for other regions, even for the world. However,
vast basin areas such as Kuwait, although little larger than the State
of Connecticut, ultimately will produce more than 80 billion bbl of oil.
Thus, such methods are not valid.

Geologists have long realized that one of the best ways to look for
oil is to explore the belts or trends in which it has already been discovered in good quantity or under similar geologic situations elsewhere.
However, not all such truths are so clearly evident.

There is a very great variation in the amount of oil that is present
in rocks of the different geologic periods, even in the different age
units of periods, and these differences are worldwide. For example,
approximately half of the oil found in the world has been found in the
sedimentary rocks of a single geologic-age span of about 50 m.y. In rocks
deposited during the rest of the 500 m.y. of Phanerozoic history, the oil
which has been discovered is distributed in much lesser percentages,
ranging down to less than 2 percent per 50 m.y. of depositional history.

Well above 90 percent of today's reserves of both oil and gas is
found in young sediments deposited within the last 35 percent of the time
since the Precambrian. Out of 62 countries that produce oil and gas,
only seven produce from Paleozoic rocks. Per-acre yields of Mesozoic-
Tertiary rocks are commonly many times more than those of Paleozoic rocks
on a worldwide basis.

UNDERSTANDING OIL OCCURRENCE

Sound petroleum resource estimates can be achieved only by correlating
the facts of oil occurrence with geology. There is no way of circumventing
this need. The reliability of such correlations depends on the degree to
which the many factors that control occurrence have been recognized and
correctly interpreted. Analyses of this sort bring recognition not only
of the extreme variability of oil and gas occurrence, but also of the
reasons for this variability.

Theories that are not based broadly and firmly on present knowledge
of oil occurrence may be more harmful than beneficial. The same is true
of criticisms that decry any efforts at quantitative resource estimates
resulting from many decades of worldwide study and analysis. Nature always follows well-regulated principles. Any apparent exceptions to those
principles (e.g., the presumed source of the oil or of time of oil migration and accumulation) generally disappear with adequate analysis, when
all facts are recognized and considered. Because estimates of ultimate
potential resources commonly differ widely, we need to examine critically
the reliability and importance of the various data and the interpretation
methods used.

Worldwide experience in studying petroleum occurrence is essential
to estimation of potential resources. There is no substitute for such
experience in analyzing the relation between oil occurrence and basin
type, architecture, and many other pertinent geologic and engineering

facts. Wide experience is certain to add immeasurably to the understanding of even the most brilliant geologist or engineer whose work has been
restricted to one region. We need not only basic knowledge, but also
balance and judgment, in order to profit from the use of exploration tools,
particularly the new glamour ideas. That an understanding of oil occurrence is still widely superficial is evidenced by the disagreements that
exist concerning some of the most fundamental principles, and also by
observation of many of the areas where money is being spent in leasing
and exploration.

OTHER METHODS OF ESTIMATING RESOURCES

Estimating how many billions of barrels of oil or cubic feet of gas
will be found at prices 20, 40, or 50 percent or more above those prevailing at the time is an exercise that has been indulged in by some,
little realizing that the nation's main fields in most areas have been
found. To what avail are higher prices if costs go up much faster? Many
of us recall the once commonly expressed view that, "when the price of oil
reaches $3.00 a barrel, shale oil will become economical."
In dealing with quantities that cannot be measured precisely, such
as potential resources, there are limits beyond which the value of additional information becomes questionable. Just as one would not measure
distances for a proposed auto trip with a theodolite, little is gained
by conjuring up questionable methods of estimation refinement. Normally,
the unit cost to market oil or gas determines whether a petroleum accumulation is of commercial or reserve caliber. As this cost rapidly becomes
more dependent on volume, many future discoveries, so optimistically forecast by some, will be found to be uneconomic or to supply relatively
little toward meeting the escalating demand.
Inasmuch as 80 percent or more of the world's petroleum reserves is
accounted for by the major fields, the amount of petroleum that is represented by commercially marginal accumulations may, because of escalating
costs, turn out to be inconsequential. Nevertheless, such additional
information as suggested by Richard Gonzalez, an economist (e.g., depths
of sands, number and size range of fields), is normally provided in each
individual basin or area being considered for investment or other purposes, including resource estimates.
Among the most important factors determining the size of both reserves
and potential resources are costs and prices. Considering the stage of
deterioration of the basic supports for world economies, however, neither
costs nor prices can today be forecast with any certainty. Gone are the
times when costs and prices remained relatively stable. Those times have
been wiped out by the rapidly escalating, economically unsupportable cost
multipliers with which our economy has been flooded over the past decades,
and by the squandering of our nation's resources.
There is hypocrisy—probably innocent—and a bit of irony in the
requests now being made by government bureaus that these factors of increasing cost be taken into account in making reserve and potential-resource estimates. Perhaps they should give us a subcategory for this!
As I view our present economic-political trend, we are playing both ends
against the middle at an accelerating rate. Considering the consequences
of this trend, we would be deluding ourselves by a proliferation of first,
second, and third categories of resources which will not be uniformly
understood, whose boundaries will constantly change, and whose meaning
will have no permanence.
Potential-resource estimates have been made on the basis of the extrapolation of production curves. Such estimates may approximate the mark

in some very exceptional instances, such as in basins or countries with
a long history of intensive exploration and production where all of the
major, easily found fields have been discovered. However, such estimates
disregard entirely the prospect that important additional reserves may
be found in the future. The Illinois basin provides an example of this
shortcoming. A forecast based on production curves for the long history
of Illinois basin production prior to 1935 would have missed the greatest
discoveries and the very major production of subsequent years.

Estimates of potential petroleum resources that are not soundly
based on geology are obviously worthless when dealing with the prospects
of undeveloped or little-developed areas. It is in these areas that
realistic estimates have their greatest value for critical investment
decisions.

EXPLORATION PRACTICES

In selection of acreage, it would appear that industry has a tendency
to overvalue greatly the less attractive and commercially submarginal
areas and to underestimate the potential of the much smaller percentage
of bonanza areas. However, many geologists are developing a more worldwide outlook with respect to the factors that control oil and gas occurrence. They are learning that the conditions and the exploration problems
they had become used to are not unique to their own areas. On the other
hand, by observing situations in many places, they are finding that some
of their most cherished views about oil occurrence and the factors that
control it have been in error. They are learning also that there is
infinitely more to petroleum geology than the "fascination of seismic
contours, shuffling electric logs, and looking at samples through binoc-
ulars."

Managements of some companies oversimplify exploration practices
and are carried away by new fads. A company may lease any sedimentary
area and then look for drilling prospects purely on the basis of surface
mapping and seismograph surveys. As most structures in the greater part
of most basins are dry, the effort may have little value—not primarily
by reason of the money spent but of the time wasted. There is also a
strong trend toward gadgetry, particularly in the fields of geophysics
and data handling. Corrections are made for weathering, normal moveout,
velocity increases, etc. Computers and data processing tools do help
speed the work, and beautiful pictures are developed. This is good but,
unless the structures and suggested traps are favorably located, they
will certainly be a disappointment. Improvement in display technique
may be made at the sacrifice of geologic interpretation. There is a
tendency to become fascinated by the processing and by the pictorial
results; the machine does our "thinking" or, perhaps, serves as an excuse
for not thinking.

Various fads have been a part of the history of oil exploration. One
is the search for stratigraphic traps, without regard for the controlling
depositional factors. Many millions have been spent to no avail exploring
for oil in updip pinchouts that were downdip at the time of deposition and
oil migration. Then, with a swing of the pendulum, the geologists are
charged with looking for reefs, most of which may be dry because of very
basic basin conditions. They may go off on another new tangent in pur-
suit of "hydrodynamic" prospects, without regard to the facts of geology,
or they may expect some miracle from electronic data processing. Today,
"bright spots," whose causes vary greatly in different basin environments,
are much in vogue.

DOUBTFUL SOURCES OF UNDISCOVERED RESOURCES

For over 4 decades, beginning back before the discovery of the East Texas oil field in 1930, geologists have eloquently evoked attention to the thousands of miles of updip stratigraphic wedges or pinchouts that are present in all basins, and have predicted that these must trap a vast amount of petroleum that will be found in the future. Stratigraphic wedging or loss of section plays some part, more or less locally, in most petroleum accumulations; however, the predominant feature not only forming the trap but having other very necessary requisites is usually other than stratigraphic. Wedgeouts have no value if they are not in the right place in the basin and were not formed at the right time geologically. Purely stratigraphically trapped accumulations can and do occur under special situations. Some will be found in the future, and a modest percentage of them will be large enough to be of real importance.

Another suggested source that has entered into our predictions of future oil far in excess of its track capabilities is the still unexplored or little-explored lower part of deep basins. I suggested many years ago that the incidence of oil occurrence must decrease below some optimum depth in nearly all basins. This depth may range between 3,000 ft (910 m) and about 10,000-12,000 ft (3,050-3,660 m), depending on the geology. There are several basic reasons for this. Principal of these is that petroleum normally migrates flankward from the deep basin areas of low porosity and hence low reservoir capacity—areas that normally are the depositional environments of muds, marls, and muddy carbonate sediments. It migrates to the higher flanks of the basin—to belts or trends of screened sands or to porous carbonate materials that accumulated in aerated environments. The optimum depth of petroleum occurrence is greater in basins of rapid deposition and low heat gradient, such as certain thick Tertiary deltas outside of mobile belts; and it normally is less in basins with a higher than average heat gradient. Of course, in addition to an optimum depth of occurrence, there is also an optimum economic depth. Nothing becomes a resource unless it can be found and produced at a profit; so small, deep accumulations would be ruled out.

Geologists, both in industry and in government, have argued that a much greater density of exploratory wells would result in the discovery of many accumulations of oil and gas—and they infer that these accumulations will provide large new reserves. These views, I believe, do not take into account the fact that most, if not all, important discoveries would almost certainly occur within the areas and trends of major accumulations, which generally have already been densely drilled. They also do not consider that about 80 percent of the world's oil and gas reserves occur in major accumulations; thus, new discoveries that are large enough to be commercial would represent a relative trickle rather than a really important addition. Also to be considered is the fact that, to date, costs of finding and producing have risen much faster than have prices.

Large volumes of potential petroleum resources have been predicted for various deep marine areas. Seemingly, these forecasts do not adequately consider important physical and environmental factors and other critical aspects of the stratigraphic history of the oceans, not to mention the economics of recovery of whatever hydrocarbons may occur.

BASIN CLASSIFICATION AS A BASIS FOR ESTIMATING RESOURCES

The so-called "typing and modeling" approach to oil-occurrence estimation, referred to by Peter Rose, Chief, Oil and Gas Resources Branch

of the U.S. Geological Survey, is an attempt to quantify occurrence in
relation to one of several common stratigraphic-tectonic settings. This
is essentially one of the lines of approach that I used in my worldwide
basin and oil-occurrence studies initiated in the 1930s. Basic to the
study is a classification of basins by manner and place of origin, history
of development, and present conditions. It seems to me that some of the
newer efforts at basin classification are unneccessarily complex and do
not recognize certain newer ideas regarding basin-development history.
For example, the variation of incidence of oil occurrence with geologic
age is of primary importance.

Although the manner and magnitude of both oil and gas occurrence vary
with basin type on an overall worldwide basis, the variations within any
one type are also very wide. In fact, there is great overlapping of the
incidence of petroleum occurrence between basin types or "models." Thus,
it may be quite impractical to use any fixed numerical yardstick based on
basin type to arrive at an estimate of the ultimate potential volume of
oil or of gas in any individual basin, or in any part or element of the
basin. Recognition of the broad variations between basin types may be
quite helpful, but it is better to consider each situation as more or
less unique. Each basin or part of a basin should be considered on its
own merits; otherwise, much wider errors than are justified may result.

I have been critical of some of the shortcomings of my profession.
I am aware of them not as a result of any unique virtue, but because I
have been interested in the worldwide occurrence of oil and, more particularly, because I have been privileged much more than most to spend my
life in a worldwide study of oil occurrence and the factors that control
it. I have discovered that I started out with a very heavy load of false
notions, and am sure that I will discover many others, for I have learned
that he who seeks will find.

FACTORS CONTROLLING OIL OCCURRENCE

The factors which control the incidence of oil occurrence are mani-
fold. Some of them are only vaguely understood; others are no doubt
unknown. Many were listed or specifically referred to in *Habitat of Oil*
(Weeks, 1958) and were discussed in somewhat greater detail in "Origin,
Migration, and Occurrence of Petroleum" (Weeks, 1961a). The following
list includes some of the factors that tend to make a basin highly prospective, or, conversely, the absence of which may make it likely to contain relatively little or no oil.

1. A favorable basin form and architecture.
2. An optimum degree of mobility during sedimentation.
3. Favorable depositional environments for both source rock and
reservoir facies, as reflected in their lithofacies and biofacies.
4. A favorable historic and physical relation between source rock
and reservoir facies.
5. A favorable ratio of reservoir rock to source rock. This factor
implies a favorable provenance, such as a granite terrane or hinterland
or other source of quartz sands. A mafic volcanic terrane is normally
worthless as a source of reservoir sands, although carbonate reefs or
reef complexes and other types of carbonate rocks may develop in such
aerated environments.
6. An adequate rate of deposition and a favorable energy balance.
7. A high degree of sorting or screening of muds from the sands.
8. Broad lateral variability in rate of deposition.

9. Large-scale lenticularity rather than uniformity of deposition across the basin. Large sand lenses or sand buildups enclosed within and/or laterally adjacent to well-developed source facies are particularly favorable.

10. The presence of unconformities, diastems, and lesser depositional breaks that are favorably positioned, in time and space, in respect to source and reservoir.

11. An optimum degree of tectonic deformation below an unconformity, particularly where overlain by good source facies.

12. Abundance, timeliness, and effective distribution in a basin of traps—structural, compactional, stratigraphic, or combinations of these.

13. The presence of depositional sinks flanked by hingebelts and/or positive structural features of tectonic or other origin.

14. Preservation from erosion of favorable relations below an unconformity.

15. Absence of fresh water flushing across the basin.

16. Protection of part or all of the original petroleum-bearing part of the basin from destruction by erosion or excessive tectonism subsequent to petroleum accumulation.

17. Direct evidence indicating that the basin sediments are petroliferous, together with evidence that the original oil is largely retained. However, an abundance of surface indications, particularly in such areas as Cuba, Morocco, and Timor, is not a quantitative index of a basin's prospects.

18. Geophysical confirmation of favorable stratigraphic and structural conditions—basin form, architecture, etc.

19. Favorable results from, and significance of, any exploratory drilling.

20. Absence of excessive metamorphism.

21. Optimum depth of prospective petroleum zones, coupled with favorable temperature gradient and temperature history.

22. Adequacy and strategic location of traps plus timeliness of trap formation. These are especially important factors.

23. Presence of evaporites (a) below the reservoir, as a favorable diapir structure (e.g., salt domes or fault blocks, salt pillows, etc), (b) above the reservoir as an excellent trap seal (as in Iran, Iraq, and Arabia), or (c) as an indicator of the probable existence of earlier or later, silled or otherwise favorable basin conditions.

24. A hydrostatic system that displays the presence of good porosity and permeability without wide artesian-flow conditions. Blanket sandstones require more structural closure and more closure permanence than do sandstone lenses. A section having a high sandstone/shale ratio tends to carry its oil lower in the basin than a section with a lower ratio.

25. A deposition rate adequate to preclude excessive bacterial decomposition of organic matter. This is a very important requisite and one which varies with bacterial oxidation.

26. Favorable age of the sedimentary units that lie within favorable basin position and burial depth.

27. Environments of deposition within a salinity range from a little above that of the normal open ocean to weakly brackish. Oil and gas generated in sediments deposited in environments with somewhat higher-than-normal salinities and a lower-than-normal pH tend to be sulfur bearing. Commonly in such environments the water temperatures are high enough for deposition of reef or other mud-free carbonate material which then forms a potentially favorable reservoir. Whether an oil is of a paraffinic, asphaltic, or some type of mixed base normally depends on the biologic as well as the sedimentary environment, and thus on whether the facies is terrestrial or marine.

28. A sediment section that was adequately thick to have placed the favorable stratigraphic zones or parts of the section within a temperature environment favorable for oil generation.

29. A favorable history of basins of the continental margin in relation to (a) their crustal position with respect to plate tectonics, (b) the *age of basin and sedimentation initiation*, (c) the provenance and type of sediments, and (d) subsequent history. These factors (taken together with the others listed) are very important in offshore regions.

In resume, other factors being equal, petroleum occurrence is related in sequence to: (1) the mechanics and nature of the basin subsidence, which vary in a very general way with basin type and basin age; (2) the resulting basin form which, along with favorable sediment supply, rate of deposition, and currents, determines sediment distribution; (3) the distribution of sediments and their relations, both of which further modify the basin form; and (4) the intrasediment environment, which is largely influenced by the basin bottom form and by the rate of deposition; and (5) the environmental factors, which determine the biofacies as well as the lithofacies distribution.

All of the foregoing factors, along with the adequacy of reservoirs, timeliness of trap formation, temperature history, and the efficacy of the seal, determine the incidence (volume and distribution) of petroleum.

DISCUSSION OF FACTORS

Although petroleum occurs as a minor constituent throughout most sediments, it is present in commercial concentrations in relatively limited places. Worldwide, the incidence of oil is decidedly greatest in those reservoir facies deposited over highs that lie within or adjacent to depressions in which the environment was favorable for accumulation of source beds.

The yield per unit of area and volume of sediment decreases toward the forelands or more stable interior areas of the continent. Beginning about 40 years ago, I contoured the per-acre yields of many basins. In basins like the Mid-Continent, the yields generally decreased up the shelf from a maximum at a belt for which I coined the term "basin hingebelt." Stable, open, aerated sea bottoms, or similar areas of basins that are remote from such favored belts or trends, have contributed very little oil. Environments that are favorable for deposition of source beds are generally unfavorable for reservoir-sediment accumulation, and vice versa. Normally, no oil can occur except where source and reservoir facies co-exist in a proper relation.

An adequate rate of deposition is an important factor in preserving organic matter from bacterial oxidation. A common fallacy is to assume that the percentage of organic matter in a sediment is a true measure of the areal incidence of preservation of organic matter. Nor is the percentage of organic matter a measure of the effective petroleum-producing ingredients of the organic matter. It is not at all a true measure of the potential of the sediment as an oil source. For example, area A may have the same percentage of organic matter as area B but, because the deposition rate was five times as great in area A, it contains five times the total preserved biomass, and probably carries it in a far more effective state.

Basin-flank areas of rapid deposition may contain as much or more organic matter per unit of time as the deep stagnant bottom areas of slow deposition where anaerobic and/or aerobic bacteria have had an opportunity to destroy much of the important organic accumulation before burial. Moreover, of all of the constituents of organic matter, the hydrocarbons

and the hydrocarbon source materials are the most easily destroyed by
bacterial oxidation, but they may be effectively preserved by rapid burial.
Various investigations of young sediments have shown that hydrocarbons
normally are found first and appear in greatest abundance in the flank
lenses of more rapid deposition.

A true source rock might be defined as one which contained organic
matter that was readily convertible to hydrocarbons and protected from
bacterial oxidation, and which had the facility for migration of the
hydrocarbons to a reservoir rock. Therefore, to evaluate source-rock
potential, it may be more important to ascertain facts of this nature
rather than percentage of organic matter.

The rate of deposition in the deep central part of a basin may be
so slow as to permit bacterial destruction of the organic matter. Inasmuch as the oxidation-reduction potential decreases rapidly below the
sediment-water interface, a rate of deposition that is relatively rapid
can be a major factor in the preservation of the principal oil source
materials and of generated hydrocarbons. Thus the best conditions for
both the source of organic matter and its preservation in many basins
may lie in the intermediate basin-flank zones where there is an optimum
balance between rate of sedimentation, temperature gradient, and other
aspects of the physico-chemical environment.

WORLD LAND AND SUBSEA RESOURCES

Sedimentary rocks underlie approximately 17 million sq mi (44 million
km^2), or about 30 percent, of the 57 million sq mi (147 million km^2) of
the earth's land area. Rough calculations suggest that, when all oil
fields on land have been found, their number may be of the order of
7,000-8,000. Of the 17 million sq mi underlain by sedimentary rocks,
it is estimated that about 20 percent, or 3.4 million sq mi (8.8 million
km^2), will eventually be found to contain oil fields. From industry
experience, it would appear that a weighted average of about 2.5 percent
of this area, or 85,000 sq mi (220,000 km^2), may be expected to be underlain by commercial oil reservoirs. Although yields range from several
thousand to several hundred thousand barrels per acre, worldwide studies
based on geology and exploration experience indicate that 35,000 bbl/acre
may be a reasonable weighted-average productivity factor. This would
result in an ultimate recovery by conventional means of 1,900 billion
bbl of oil.

The foregoing was *not* the way my various estimates of ultimate potential world oil reserves, recoverable by conventional production means
from the lands, were obtained. It merely approximates some weighted
averages made by working backward.

Worldwide gas consumption is about 120 Bcf/day, equivalent in Btu
value to about 10 million bbl/day of crude oil. This is 36 percent of
the present crude oil consumption of about 56 million bbl/day. The
United States accounts for over 60 percent of the world's consumption
of gas. Total world reserves of gas today must be close to 2,000 Tcf.
About 80 percent of proved world oil reserves is accounted for by major
or giant ·fields, ranging from 0.5 billion to 80 billion bbl. If the
trend of recent decades could continue, over the next 20 years the world
demand would be something like 600 billion bbl of oil and 1,100 Tcf of
natural gas.

Considerable study of the continental shelves has been carried out
since the early 1940s. Combining the extensive geologic knowledge and
exploration experience on land with that accumulating for offshore areas,
and applying them to the approximately 20 million sq mi (52 million km^2)

of shelves and slopes, I have arrived at ultimate potential oil resources for each of the individual offshore basin areas. These are summarized in Table 1.

UNITED NATIONS' ACTIVITY IN RESPECT TO RESOURCES

Early in 1973, the United Nations requested my assessment of the total ultimate subsea petroleum resources of the world, divided between 10 major subdivisions of the world's oceans. It was requested further that in each subdivision the assessment be divided landward and seaward of each of four boundaries that have been most commonly discussed and/or demanded by the nations of the world to be established between national and international domain. The purpose of the assessment was to serve as a basis (no doubt just one of many proposed bases) for the UN deliberations on division of ownership of these resources. These deliberations have been proceeding for 16 years. In response, I set forth my estimates in a 25-page document, together with 16 tables and maps. A revised version of this was presented to the Southwestern Legal Foundation and later published (Weeks, 1974b).

The subsea petroleum estimates contained in my 1973 report to the United Nations were somewhat greater than those given in Table 1, for they also included the oil Btu equivalent of the gas, amounting perhaps to fully half of that estimated for the liquids. Also, they included relatively minor, somewhat doubtfully economic amounts from various deep-bottom physiographic features. Actually, an error of 50 percent too low or 100 percent too high in any well-based estimate will make little, if any, important difference in the time when production of petroleum will reach its maximum, or in the total length of petroleum supply. The timing will be determined by (1) the rapidly escalating growth in world demand for energy, (2) the finite nature of potential petroleum resources, and (3) the fact that petroleum is and will continue to be, up to the limit of its availability, both the preferred fuel for all purposes and an unmatched raw material for a vast number of uses.

On December 6, 1973, the United Nations, without a dissenting vote, agreed to the establishment of a United Nations University. Emphasis is to be placed on research in the resolution of problems of most critical interest to the nations of the world. Foremost among these concerns is the shortage of energy. Japan immediately subscribed $100 million toward the University and also agreed to erect a central core of university buildings north of Tokyo. Other nations are considering similar centers for research on energy sources as well as on other major problems.

In April 1974, representatives of government, industry, universities, and the United Nations met in the United States to discuss what needs to be done in the implementation of remedial research. The U.S. National Committee (under the Department of the Interior) held two additional meetings in 1974, the last of which was in association with the 9th World Energy Conference in Detroit. Other meetings are planned.

100-YEAR ENERGY OUTLOOK

The part that each energy source would logically play in supplying the projected energy demand over the next 100 years was published in 1959 on the occasion of the oil industry's first centennial, and there have been various revisions (Weeks, 1961b). A graph of the projection for the world is shown in Figure 1. The finite nature of petroleum resources is shown clearly against the background of the projected total

TABLE 1. ESTIMATED PRODUCTIVE AREA AND POTENTIAL OIL RESOURCES OF SUBSEA SHELVES AND SLOPES

Area	(Sq Mi)	(Km²)	% of Total
Total area	20,000,000	51,800,000	100
Total nonbasin area	8,000,000	20,720,000	40
Total basin area	12,000,000	31,080,000	60
Commercially submarginal	8,000,000	20,720,000	40
Commercially attractive	4,000,000	10,360,000	20
Bonanza class (20%)	800,000	2,072,000	4
Moderately attractive (80%)	3,200,000	8,288,000	16

Computation of Potential Oil Resources

Bonanza class   800,000 sq mi X 2.5 percent productive = 20,000 sq mi (51,800 km²) or 12,800,000 acres

Moderately attractive   3,200,000 sq mi X 1.5 percent productive = 48,000 sq mi (124,320 km²) or 30,720,000 acres

12,800,000 acres X 40,000 bbl/acre = 512,000,000,000 bbl

30,720,000 acres X 25,000 bbl/acre = 767,000,000,000 bbl

Ultimate recoverable liquids   1,280,000,000,000 bbl

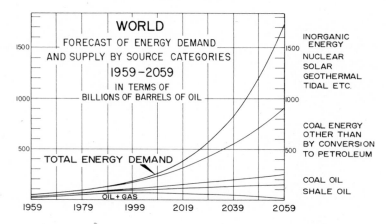

FIG. 1—World forecast of energy demand and supply, 1959-2059, based on demand trend as of 1959.

energy demand over the next 100 years. The portrayal on Figure 1 assumes that energy from the various sources can be made available economically.

Organic sources include oil and gas, shale oil, heavy-oil sands (erroneously called "tar sands"), coal, wood, and farm wastes; the inorganic sources include water power and nuclear, solar, and geothermal energy. Today, wood and farm wastes, once the predominant source of world energy, still provide as much energy as do atomic sources.

Projection of the curve of total energy demand was based on the assumption that the upward trend of energy demand until 1959 would continue throughout the next century. The estimated quantitative schedule of use of each of the potentially available energy sources is portrayed graphically beneath the curve of total energy use.

Because of the various factors that are not clearly predictable, projection curves of total energy demand and supply cannot be made confidently beyond the near future. Nevertheless, the exercise of showing how the various potential sources might logically supply the projected total energy use (based on their indicated total availability and economics of supply) was in some respects very revealing. For instance, it shows that petroleum from conventional sources will supply a gradually lessening percentage of the total energy needs over the next decade or two, and that thereafter it will supply a much more rapidly decreasing, in fact a very minor, percentage of the projected energy demand. By the end of the 20th century, petroleum from conventional sources will supply less than one third of the projected energy demand both of the United States and the world, and, by the year 2020, this will have decreased to less than one fifth (the assumption is made that the projected sources of total energy will be economically feasible).

The charted curves of Figure 1 were prepared in the late 1950s, before the spurt in energy use of the last 15 years. A similar chart, for 120 years from 1960 to 2080, was prepared early in 1974 (Fig. 2). The growth in energy demand increased more than 100 percent per decade between 1950 and 1970. Because of the uncertain state of world economics, this rate was reduced to 75 percent for the 1980s, then to 50 percent for the 1990s. For projections after the year 2000, I have progressively lowered the rate of increase of total energy demand to 20 percent per decade on one curve and to 10 percent per decade on another. The figures from which the three curves were constructed are shown in Table 2.

Regardless of the rate of increase of total world energy demand, short of a general worldwide economic collapse, the proportion of total demand that conventional oil and gas is destined to supply is relatively small. This would be true even if the already rather optimistic amount of oil and gas resources shown were increased 100 percent, or if there were considerable additional production from secondary recovery.

In respect to the energy shortage the cry goes up, "Why have we not been told?" The fact is, they have been told repeatedly over the past 30 years or more. They have even been given the numbers, many times. By actual count, I have published such data no less than a dozen times since 1940, and announced them widely in lectures. They were presented at UN Headquarters in 1949 when I served as a delegate to the United Nations Conference on the Conservation and Utilization of Natural Resources. At that conference, and on many other occasions before and since 1949, attention was called, by means of the potential resource estimates themselves, to the very finite nature of the petroleum age and to the inevitability of inadequate supply, a fact which we perhaps now, belatedly, can appreciate.

World conventional oil production probably will reach its maximum late in this century or early in the 21st century. However, it will fail to fulfill demand of the current very excessive kind long before that. Indeed, we already see the beginning of this shortfall. After reaching its maximum, production will begin a decline lasting for well over a hundred years, contributing smaller and smaller fractions of the world energy needs. The supply eventually will become a relative trickle, but petroleum still will be in demand. Thus, a large portion of the potential conventional petroleum resources of today will occupy a gradually lessening percentage of the space beneath the curves of total demand that I have projected.

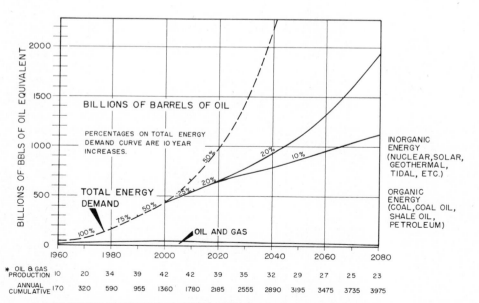

| ★ OIL & GAS PRODUCTION | 10 | 20 | 34 | 39 | 42 | 42 | 39 | 35 | 32 | 29 | 27 | 25 | 23 |
| ANNUAL CUMULATIVE | 170 | 320 | 590 | 955 | 1360 | 1780 | 2185 | 2555 | 2890 | 3195 | 3475 | 3735 | 3975 |

★ GAS IS DESTINED TO MAKE UP AN INCREASINGLY GREATER PERCENTAGE OF TOTAL PRODUCTION DURING THE 120 YEAR PERIOD. (6000 CF OF GAS = 1 BBL OF OIL)

FIG. 2—World forecast of energy demand, 1960-2080, based on figures given in Table 2.

TABLE 2. WORLD TOTAL ENERGY DEMAND, 1960-2080[1]
(IN BILLIONS OF BARRELS OF OIL[2])

End of Decade	First Assumption			Second Assumption			Third Assumption		
	% Increase Per Decade	Demand Annual	Demand Cumulative	% Increase Per Decade	Demand Annual	Demand Cumulative	% Increase Per Decade	Demand Annual	Demand Cumulative
1960		40	2,500[3]		40	2,500[3]		40	2,500[3]
1970	100	80	3,100	100	80	3,100	100	80	3,100
1980	100	160	4,300	100	160	4,300	100	160	4,300
1990	75	280	6,500	75	280	6,500	75	280	6,500
2000	50	420	10,000	50	420	10,000	50	420	10,000
2010	50	630	15,250	25	525	14,725	25	525	14,725
2020	50	945	23,125	20	630	20,500	20	630	20,500
2030	50	1,418	34,940	20	756	27,430	10	693	27,175
2040	50	2,127	52,665	20	907	35,745	10	752	34,400
2050	50	3,190	79,250	20	1,088	45,501	10	827	42,295
2060	50	4,785	119,125	20	1,305	57,466	10	910	50,980
2070	50	7,178	178,940	20	1,566	71,821	10	1,000	60,530
2080	50	10,767	268,715	20	1,880	89,051	10	1,100	72,030

[1] Assuming 10-year percentage increases as indicated.
[2] Gas computed as Btu equivalent.
[3] Approximate order of magnitude.

FACTORS AFFECTING OUTLOOK AND RESOURCE PREDICTIONS

In making projections of future energy use it is necessary to make certain assumptions, the future fulfillment of which many may logically question today. We may assume that there will be no disastrous or catastrophic events of natural origin, or of man's making, that will materially lower the present trend of energy demand. On the other hand, we may assume that acceleration of demand will not be significantly greater than that of today. This brings us to another basic assumption—that there will be no limit to man's ability to find a way to utilize an existing potential resource as long as the economic incentive and political climate are favorable. Therein lie sound reasons for doubt today.

Also affecting the outlook today are the imminent shortages of many other basic raw materials. In all of them, there is major dependence on the energy supplied by petroleum. Not the least serious among these shortages are those of food. All are aggravated by unsupportable costs born of greed, parasitism, wars, overpopulation, etc. Shortages, of course, mean higher prices, and thus are one of the causes of inflation. As these factors are not clearly predictable, projection curves of both demand and supply for any more than the immediate future cannot be confidently drawn on the basis of recent historic trends.

Half of all of the oil ever produced has been taken from the earth since 1964. This gives a mental picture of the escalating growth in world demand. What the trend portends should be evident when visualized in respect to any rational assessment of our finite potential resources. Europe's petroleum needs have more than quintupled since 1950. In Europe and Japan the consumption of energy has been growing at three to five times the rate of population increase. It is predicted that the giant new fields of the North Sea will produce 2-3 million bbl/day by 1980 and perhaps 5 or 6 million bbl/day by 1990, but at the current rate of consumption growth these large local supplies will fill little more than meet the increase in potential demand of the adjacent North Atlantic countries, and will only satisfy that demand temporarily. Fully half of Europe's supply will still have to come from other sources, such as the Arab countries, assuming that they will continue to make it available in amounts adequate to meet the escalating demand.

It should be remembered also that even in the Organization of Petroleum Exporting Countries (OPEC), as in most of the producing basins of the world, most of the oil already has been found. The ruling Arabs and others of the OPEC group know that better than our people do, and this is a main reason for their tightening up on oil export and also for the great increase in its price. So great is the accelerating demand for energy that the years of abundant supply at supportable prices are relatively few, and this, perhaps more than anything else, will force a diminution of petroleum use.

MISCONCEPTIONS CONCERNING ALTERNATIVE ENERGY SOURCES

A common remark is, "When our petroleum is gone, we will just get our energy from coal, oil shale, nuclear, geothermal, solar, or some other energy source." We will—but in greatly restricted amounts and at much higher prices, and only after 2 or 3 decades of waiting for the necessary research. Because of the very low prices, in comparison with their utility, at which all forms of petroleum energy have sold for many decades, there has been no incentive to develop other sources of energy. The historically low price/earnings ratios of oil and gas company shares are a reflection of the contribution that the industry has continuously given to the nation's economic progress.

We are seriously unprepared for the use of other potential energy sources. Today, only a minor percentage of the potential power of the uranium atom is being used, and unless and until the potentially hazardous breeder reactor is perfected, uranium will soon be in short supply also. At higher prices, appreciable amounts of lower grade uranium ore would become commercial if it were not for the fact that costs rise much faster than prices. Unsupportably high costs are a measure of the artificial support we give to our so-called "high standard of living," selfishly leaving the inevitable fateful reckoning to our descendants as an awful burden on the future of our country.

Most people generally do not realize that energy sources other than petroleum are convenient only for such uses as the development of electricity and heating, not for the major energy uses of ground and air transportation. Power based on electricity apparently is our only prospect for the use of inorganic fuel sources. For two or more decades, the same will also be true for such fossil fuels as coal and the various forms of pseudo-petroleum that can be fashioned from coal, shale oil, and heavy-oil sands. Electricity, unfortunately, is one of the most wasteful, as well as one of the most localized, uses of primary energy.

Petroleum is by far the most efficient and versatile of energy sources. It is the only raw material for the truly vast array of petrochemicals we depend on every day for many hundreds of products—from beauty creams to clothing, electrical insulation, plastics, auto tires, construction materials, etc. It is unique by reason of this vast utility, versatility, and, above all, its very low cost by comparison with all other energy sources. If this were not true, the world long ago would have concentrated on one or more substitute sources.

Just as we have wasted our precious energy heritage, we continue to throw away billions of dollars each year on an escalating profusion of wants, pretended needs, and symptoms. It is well past time that we, as a nation and individually, begin to mature and to forget the symptoms by recognizing, admitting, and eliminating their causes. National policy can rise no higher than the standards of the people.

SELECTED REFERENCES

Weeks, L. G., 1948, Basin development, sedimentation and oil occurrence: New York, Standard Oil Co. of New Jersey (now Exxon).

——1950a, A discussion of potential oil reserves: United Nations Conference on Conservation and Utilization of Resources, 1949, Proc., Plenary Sessions, v. 1, p. 107-110.

——1950b, Concerning estimates of potential oil reserves: AAPG Bull., v. 34, no. 10, p. 1947-1953.

——1952, Factors of sedimentary basin development that control oil occurrence: AAPG Bull., v. 36, no. 11, p. 2071-2124.

——ed., 1958, Habitat of oil—a symposium: AAPG, 1384 p.

——1959, Geologic architecture of circum-Pacific: AAPG Bull., v. 43, no. 2, p. 350-380.

——1961a, Origin, migration, and occurrence of petroleum, Chap. 5 *in* G. Moody, ed., Petroleum exploration handbook: New York, McGraw-Hill, p. 5-1 - 5-50.

——1961b, The next hundred years energy demand and sources of supply: Jour. Alberta Soc. Petroleum Geologists, v. 9, no. 5, p. 141-157.

——1962, World gas reserves, production, occurrence, *in* Economics of the gas industry, v. 1: New York, Matthew Bender & Co., p. 71-132.

——1964, World petroleum exploration review: 6th World Petroleum Cong., Frankfort, Germany (1963), v. 1, p. 231-271.

——1966, Assessment of the world's offshore petroleum resources and exploration review, *in* Exploration and economics of the petroleum industry, v. 4: New York, Matthew Bender & Co., p. 115-148.

——1967, Marine geology—economic problems and prospects: New York Acad. Sci. Annals, v. 136, art. 20, p. 549-574.

——1968, Australian offshore exploration—novel handling of a successful major program, *in* Exploration and economics of the petroleum industry, v. 6: New York, Matthew Bender & Co., p. 89-122.

——1971, Marine geology and petroleum resources: 8th World Petroleum Cong. Proc., v. 2, p. 99-106.

——1972, Critical interrelated geologic, economic, and political problems facing the geologist, petroleum industry, and nation: AAPG Bull., v. 56, no. 10, p. 1919-1930.

——1973, Subsea petroleum resources in relation to proposed national-international jurisdiction boundaries: New York, United Nations.

——1974a, Australia in the energy crisis: IPA Review, v. 28, no. 1.

——1974b, Subsea petroleum resources in relation to proposed national-international jurisdiction boundaries and imminent energy problems, *in* Exploration and economics of the petroleum industry, v. 12: New York, Matthew Bender & Co., p. 91-132.

——1974c, The energy crisis is real: Wisconsin Alumnus, v. 75, no. 5.

Basin Consanguinity in Petroleum Resource Estimation [1]

J. W. PORTER [2] and R. G. McCROSSAN [3]

ABSTRACT   Any technique for estimating petroleum resources, no matter how detailed, should have a framework within which the data, concepts, and principles related to petroleum occurrence can be organized. The sedimentary basin is an entity that can be identified with a minimum amount of information early in the exploration history of a region, and that is capable of yielding a very tangible indication of hydrocarbon potential and its mode of occurrence.

There is no single ideal method of estimating resource potential. The method selected should be that best suited to the purpose of the study and to the technical resources and data bases available. Thus, for a worldwide assessment of resources including studies of unexplored regions or basins, one must select a method based on higher order characteristics before attempting any detailed approach.

In classifying the Phanerozoic Canadian basins for purposes of estimating petroleum resources, it was found that their evolution in both time and place was orderly relative to generally accepted principles of continental drift. There is clearly an evolution in basin styles through time, and there are corresponding, distinctive families of trapping configurations in each basin class.

The basin classification is placed in a chronogenetic framework through study of basin evolution in four major stratigraphic slices called "megasequences." An examination of giant oil and gas fields of the world within this basin classification framework demonstrates distinctive modes of occurrence in time (megasequence) and space (crustal position).

INTRODUCTION

A broad regional framework within which petroleum estimates can be organized is desirable regardless of the detail of the estimates. Ideally, the best estimate will be made using all information available on every prospect and "play" in a region—including the exploration history. The estimate should be made by examining each critical parameter in a probabilistic framework and by ultimately combining these data to provide a total estimate for the region, basin, etc. Unfortunately, this ideal is not always attainable, because of the lack of either specific information or of the time and resources to make complete estimates.

Evaluations of whole basins or large regions on the basis of regional geology codified within a basin classification scheme offer an alternative and are an essential prerequisite to the use of volumetric data. Petroleum yield factors should not be used indiscriminately over large regions but, rather, should be derived for, and applied to, specific basin types, preferably as ranges. Even where the detailed approach is feasible, moving from the general to the particular, or from regional to local scale, seems a logical approach. In evaluating the petroleum resources of a continent or of the world, one would hardly start by evaluating every prospect in detail. Detailed studies must be consistent with the regional framework; therefore, it is necessary to classify larger units of regional geology into categories corresponding to petroleum potential and mode of occurrence.

The sedimentary basin seems to us to be the most logical building block for such study. This unit groups together depositional and structural characteristics such as source, seal, reservoir, and timing, which

[1] Manuscript received, March 3, 1975.
[2] Canadian Superior Oil Ltd., Calgary, Alberta.
[3] Geological Survey of Canada, Calgary, Alberta.

bear on the generation, migration, and entrapment of petroleum, and thereby provides a first rough screen for examination of petroleum potential.
It is obvious that certain basin types are much more prolific for oil and
gas production.

BASIN CLASSIFICATION

The sedimentary basin should be used as the basis for petroleum estimates because it is an entity that can be established with a minimum
amount of information early in the exploration history of a region, and
an entity capable of yielding a very tangible indication of hydrocarbon
potential and its mode of occurrence. For example, a few reconnaissance
seismic profiles are sufficient to ascertain the basin style and can be
expected to yield indications of the probable trapping mechanisms and
configurations. The classification of the basin immediately sets up a
model within which one can hypothesize a limited number of petroleum
systems.

The sedimentary basin may then be examined in detail for the properties of importance in petroleum accumulation and rated within the basin
classification. A check list, including 16 main headings, was used to
examine the Canadian basins. In examining the Canadian basins, we found
several basins of poor quality which generally lacked more than one key
factor.

There are many excellent basin classification schemes in the literature designed to serve various purposes. Although the one developed by
Klemme (1971) most closely fitted our needs of categorizing petroleum
occurrence of the Canadian basins, we made certain modifications to it
that seemed helpful in our work.

The fundamental control determining basin character is crustal sta-
bility. Some topographic relief, generally of structural origin, must
exist to act as a trap that can accumulate sufficient sediment to initiate
isostatic movements which, in turn, can enhance the structural accomoda-
tion. Though it may seem tautological, some definition of the term
"basin" seems desirable in view of variations in usage over the years.
A *sedimentary basin*, as used in this discussion, is an irregular surface
generally of tectonic origin and its contained mass of rock formed from
sediment derived from one or more provenances and originally deposited
in a relatively uniform tectonic environment within a discrete geographic
area over some reasonably long period of geologic time. This definition
corresponds more or less to current usage even though it may appear illog-
ical that the term "basin" refers to the sedimentary content rather than
the topographic depression. The term may thus include the deposits of
an unconfined topographic depression such as a coastal-margin basin.

The forces that created the basin provide a certain structural style
and encourage a particular sedimentologic response. These forces may
continue to act as deposition continues and even after it terminates.
Sedimentation itself may enhance the structural activity through isostatic
and more superficial gravitational adjustment. The basin may be termi-
nated by a major orogenic event and may be succeeded by another basin.
Subsequent deformation of a basin may be characteristic of a certain
basin style, and perhaps should be considered as a logical modification
whereby the disturbed area is treated as a basin subdivision, especially
if it has particular modes of petroleum occurrence. The structural basin
merely represents a subsequently deformed sedimentary basin and represents
the final stage in its evolution. It is the interaction of tectonic
events and sedimentation in a particular place over a period of time
that produces a particular assemblage of petroleum occurrences. The

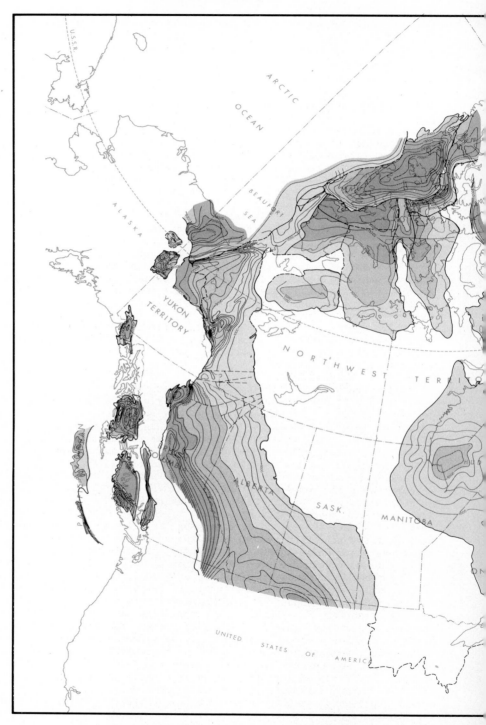

FIG. 1—Unmetamorphosed

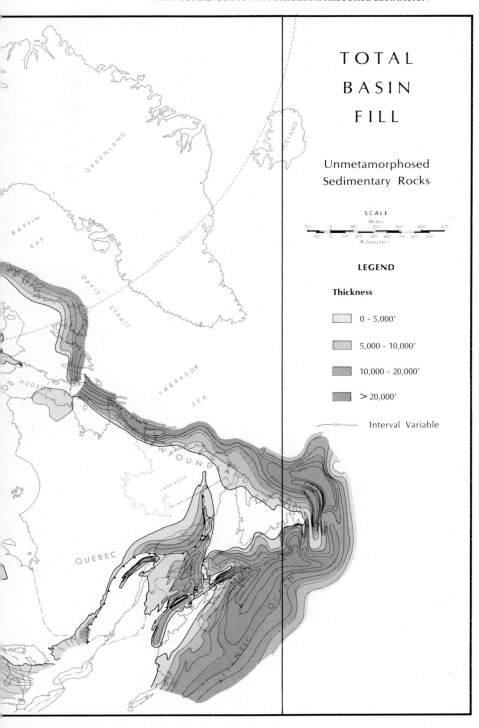

ry rocks in Canadian basins.

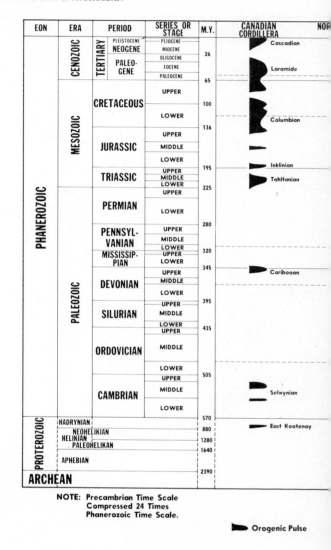

EON	ERA	PERIOD	SERIES OR STAGE	M.Y.	CANADIAN CORDILLERA	NOR
PHANEROZOIC	CENOZOIC	TERTIARY NEOGENE	PLEISTOCENE		Cascadian	
			PLIOCENE			
			MIOCENE	26		
		PALEO-GENE	OLIGOCENE		Laramide	
			EOCENE			
			PALEOCENE	65		
	MESOZOIC	CRETACEOUS	UPPER	100		
			LOWER	136	Columbian	
		JURASSIC	UPPER			
			MIDDLE			
			LOWER	195	Inklinian	
		TRIASSIC	UPPER		Tahltanian	
			MIDDLE			
			LOWER	225		
	PALEOZOIC	PERMIAN	UPPER			
			LOWER	280		
		PENNSYL-VANIAN	UPPER			
			MIDDLE			
			LOWER	320		
		MISSISSIP-PIAN	UPPER			
			LOWER	345		
		DEVONIAN	UPPER		Caribooan	
			MIDDLE			
			LOWER	395		
		SILURIAN	UPPER			
			MIDDLE			
			LOWER	435		
		ORDOVICIAN	UPPER			
			MIDDLE			
			LOWER	505		
		CAMBRIAN	UPPER		Selwynian	
			MIDDLE			
			LOWER	570		
PROTEROZOIC		HADRYNIAN		880	East Kootenay	
		NEOHELIKIAN HELIKIAN PALEOHELIKAN		1280 1640		
		APHEBIAN		2390		
ARCHEAN						

NOTE: Precambrian Time Scale
Compressed 24 Times
Phanerozoic Time Scale.

▶ Orogenic Pulse

FIG. 2—Chronology of principal tectonic elements of northern No

underlying tectonic activity which formed the basins and, ultimately, the trapping configurations may range from the opening of an ocean basin to the gentle warping of the craton.

CANADIAN BASINS

The writers, in summarizing the geology and petroleum potential of the Canadian basins (McCrossan and Porter, 1973), found that the basin evolution seemed orderly in time, place, and style relative to generally accepted principles of seafloor spreading. Before attempting to examine the general relevance of these basin families to worldwide petroleum

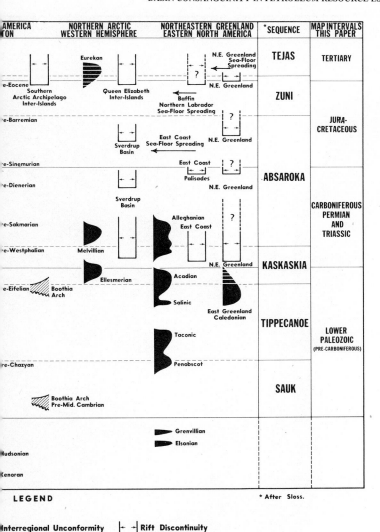

AMERICA ON	NORTHERN ARCTIC WESTERN HEMISPHERE	NORTHEASTERN GREENLAND EASTERN NORTH AMERICA	*SEQUENCE	MAP INTERVALS THIS PAPER
	Eurekan	N.E. Greenland Sea-Floor Spreading ?	TEJAS	TERTIARY
e-Eocene Southern Arctic Archipelago Inter-Islands	Queen Elizabeth Inter-Islands	N.E. Greenland Baffin Northern Labrador Sea-Floor Spreading	ZUNI	JURA- CRETACEOUS
e-Barremian		?		
	Sverdrup Basin	East Coast Sea-Floor Spreading N.E. Greenland		
e-Sinemurian		East Coast ?	ABSAROKA	
e-Dienerian		Palisades N.E. Greenland		CARBONIFEROUS PERMIAN AND TRIASSIC
	Sverdrup Basin			
e-Sakmarian		Alleghanian ? East Coast		
e-Westphalian	Melvillian			
	Ellesmerian	N.E. Greenland	KASKASKIA	
e-Eifelian Boothia Arch		Acadian		
		Salinic		
		East Greenland Caledonian	TIPPECANOE	LOWER PALEOZOIC (PRE-CARBONIFEROUS)
		Taconic		
e-Chazyan		Penobscot		
	Boothia Arch Pre-Mid. Cambrian		SAUK	
		Grenvillian		
		Elsonian		
Hudsonian				
Kenoran				

* After Sloss.

LEGEND

ica with megasequences used in this paper shown in right column.

occurrence, a very brief sketch of the types is necessary. Figure 1 is an isopach map of the total basin fill of the relatively undeformed sedimentary rock of Canada. The darker shades represent the thicker deposits, which range to more than 30,000 ft (9,144 m) in thickness. About 30 named basins and subbasins were separated on the basis of relative thickness from within the areas of basin types shown on Figure 3. In reality, of course, these grade laterally into one another and may be succeeded vertically by subsequent basins.

The regional geology was studied in four major time-rock units called "megasequences" (Fig. 2), which are separated on the basis of distinct orogenic events responsible for basin formation. These megasequences appear to have worldwide significance and seem to conform to major events

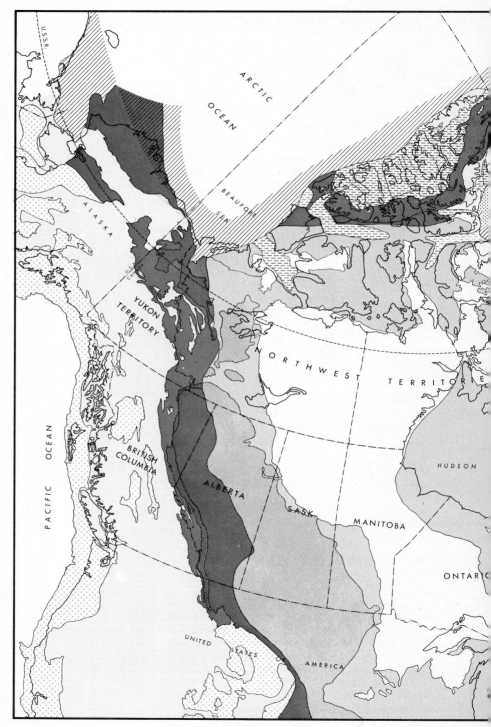

FIG. 3—Basin classific

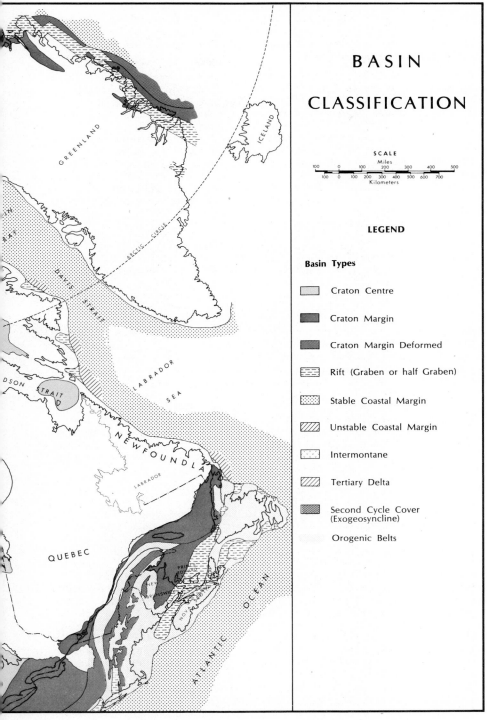

BASIN

CLASSIFICATION

SCALE
Miles
100   0   100   200   300   400   500
100 0 100 200 300 400 500 600 700
Kilometers

LEGEND

Basin Types

Craton Centre

Craton Margin

Craton Margin Deformed

Rift (Graben or half Graben)

Stable Coastal Margin

Unstable Coastal Margin

Intermontane

Tertiary Delta

Second Cycle Cover
(Exogeosyncline)

Orogenic Belts

ern North America.

preceding and following the breakup of Pangaea. In Canada, the preserved record of each of the four megasequences is characterized by the successive introduction of dominant tectonic styles and the consequent initiation of the following basin forms:

1. Stable, low-lying continents with widespread carbonate and evaporite deposits. The stable interiors are characterized by relatively undeformed, wide shallow basins. Early Paleozoic basins of other types probably formed around the continental margins but became incorporated into the surrounding orogenic belts and were destroyed.

2. Rift or collapse basins, containing widespread evaporitic and subsequent cyclic deposits with associated mafic igneous intrusions. These basins formed as a result of tensional conditions on and near the continental margins and generally were floored by old cratonic basement.

3. Intermontane basins with dumped sediment in the west where compressional tectonism prevailed and, on the eastern seaboard, stable coastal-margin basins.

4. High-standing continents characterized by delta formation.

Hence, there was a grand Phanerozoic cycle beginning with low-lying stable continents with little remaining record of extracratonic deposition and culminating in high-standing continents with predominantly extracontinental deposits.

For each megasequence two maps were prepared; one shows paleogeology and tectonics and the other lithology and thickness (McCrossan and Porter, 1973). The first pair in the series, for example, represents the lower Paleozoic megasequence. Unfortunately, these maps cannot be reproduced without the use of color. They, and the subsequent maps in the series, illustrate the gross tectonic events and their sedimentologic response during the time represented by each megasequence. The maps illustrate the different modes of basin formation that have evolved throughout Phanerozoic time.

The regional geology indicates major depositional changes that reflected the evolution of the various basin styles in time. These are shown on the basin-classification map (Fig. 3), and the time of development is indicated on the chart (Fig. 4). Even the gross basin geometry illustrated by the profiles on the chart is very characteristic for each family.

Lower Paleozoic Megasequence

Two basin types of the stable cratonic environment began their evolution during the time represented by the lower Paleozoic megasequence. The two types recognized here have distinctly different petroleum potentials, and for that reason alone are worth distinguishing. The first type we call "craton center" (CC).

The basins lying toward the craton center are for the most part filled with lower Paleozoic rocks deposited on the Precambrian basement during times of maximum transgression. Because this central part of the craton was the most positive area during all of Phanerozoic time, the basins subsided very little and are typically shallow, saucerlike depressions with very thin, widespread carbonate stratigraphic units which have been eroded extensively at various times. This characteristic geometry is expressed in a profile shown in Figure 4. The basins are separated from one another or divided into subbasins by broad regional arches. Small- to medium-size, stratigraphically trapped accumulations of oil with very little gas are the most likely targets of hydrocarbon exploration. The CC basins and each of the other families of basins have a

characteristic geology related to the structural setting and a distinct
group of trapping configurations. These features have been described
elsewhere in more detail (McCrossan and Porter, 1973).

The second type of basin style is called "craton margin" (CM) and
lies peripheral to the craton center. This basin style is represented
by the thickened wedge of the foreland deposits. The section becomes
thicker and increasingly complete stratigraphically outward from the
craton center. Like the CC basins, the CM basins are relatively stable.
This basin style grades into the CC basin type where no arch is present
to isolate the basins geographically. In Western Canada, these two types
are separated at the 7,000-ft (2,134 m) basin-fill isopach. Whether
this thinner part of the wedge is considered a basin type or a subdivision, it should be recognized as an entity having the properties of the
CC type. The characteristics of the two basin types lying on either
side of this dividing line are quite different in aggregate, as are
the quantities of discovered hydrocarbons and the future potentials.

In the CM basins, much greater thicknesses of sediment accumulated,
and thus more strata are preserved, than in the more central areas of
the craton. These rocks are much less shelflike in character and mani-
fest a greater potential for hydrocarbon generation and better sealing
characteristics for the traps. The depth of burial in the CM basin of
Western Canada is of an order that ensures the full spectrum of hydrocarbon products ranging from immature biogenic dry gas through oil to
dry gas from mature organic facies. The average ratio of gas to oil is
slightly below 6,000 cu ft/bbl, which is about the average for North
America. Almost all the Canadian petroleum reserves are found in stratigraphic traps in this type of basin.

The "craton margin disturbed" (CMD) category is not really a basin
type, but is a tectonic subdivision of the CM basin. It is worth singling
out, however, because its hydrocarbon potential and its structural and
sedimentary characteristics are quite different from those of the two
preceding basin categories. It is simply a deepening of the CM basin
into the fringe of the orogenic belt where it has been affected by major
compressional deformation.

Upper Paleozoic – Triassic Megasequence

In this megasequence the rift or extensional (R) basin becomes important. This structural style, present along the eastern and northern
part of North America, is a modification of the craton margin. This
megasequence represents a period of rifting or collapse which presumably
presaged the breakup of Pangaea. Compared with the previously discussed
basin types, the rift basins were formed and filled in a relatively short
space of geologic time (see Fig. 2), and, as might be expected, the important trapping configurations are related to tensional structures.

All of these basins have a tendency toward accumulations of gas
rather than oil because of the great depths of burial and probably higher-
than-normal geothermal gradients at the time of their formation and filling.

Jurassic-Cretaceous Megasequence

During the period of this megasequence (shown on illustrations as
"Jura-Cretaceous Megasequence"), three additional basin styles began to
evolve. Perhaps the most widespread and obvious are the "stable coastal-
margin" (SCM) basins of Eastern Canada. The profiles of the stable
coastal-margin basins characteristically are quite thick but broad and
lenticular, and the thickest sections lie beneath the slopes.

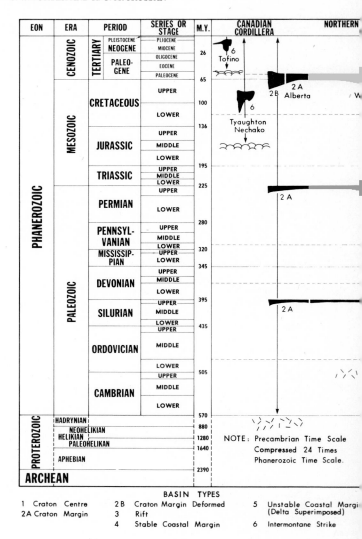

EON	ERA	PERIOD	SERIES OR STAGE	M.Y.	CANADIAN CORDILLERA	NORTHERN

FIG. 4—Time of formation of Canadian basin ty

BASIN TYPES

1 Craton Centre	2 B Craton Margin Deformed	5 Unstable Coastal Margi (Delta Superimposed)
2A Craton Margin	3 Rift	
	4 Stable Coastal Margin	6 Intermontane Strike

Basins of the SCM type stretch along the eastern seaboard of North America. In the southern part of the belt, the initial deposits formed were evaporitic and seem to reflect the restricted nature of the proto-Atlantic. These beds are succeeded by a great thickness of predominantly clastic deposits, sandier toward the shore, which extend throughout the Jurassic into the Tertiary sequence. The lithology of these rocks suggests an environment unsuitable for the preservation of abundant organic material.

The geometry of the "unstable coastal-margin" (UCM) basin is characterized by the occurrence of the maximum thickness beneath the shelf. A category for structurally more active, extensional continental margins was useful to represent areas of more effective sediment trapping.

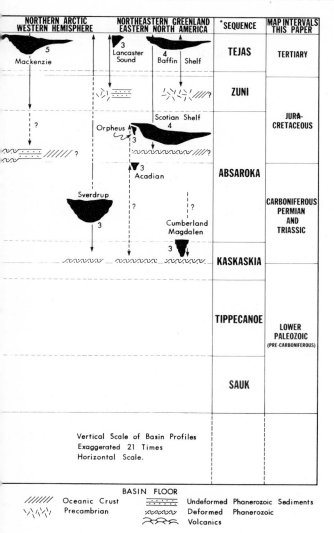

NORTHERN ARCTIC WESTERN HEMISPHERE	NORTHEASTERN GREENLAND EASTERN NORTH AMERICA	*SEQUENCE	MAP INTERVALS THIS PAPER
5 Mackenzie	3 Lancaster Sound 4 Baffin | Shelf	TEJAS	TERTIARY
		ZUNI	
?	Orpheus 3 Scotian Shelf 4		JURA-CRETACEOUS
?	3 Acadian	ABSAROKA	
Sverdrup 3	? ? Cumberland Magdalen 3		CARBONIFEROUS PERMIAN AND TRIASSIC
		KASKASKIA	
		TIPPECANOE	LOWER PALEOZOIC (PRE-CARBONIFEROUS)
		SAUK	

Vertical Scale of Basin Profiles
Exaggerated 21 Times
Horizontal Scale.

BASIN FLOOR
////// Oceanic Crust   Undeformed Phanerozoic Sediments
\/\/\/ Precambrian   ~~~~ Deformed Phanerozoic
           ZZZ Volcanics

ed examples are shown as traverse profiles.

For the UCM basins, it is suggested that a collapsed segment of stable coastal margin is created where down-to-the-ocean faults caused by strong, negative movements are intersected by other fault systems such as transform complexes. Rainwater (1972) reasoned that such faults could have channeled regional drainage into these areas, resulting in the rapid deposition of great thicknesses of organic-rich sediments with facies relations and environments favorable for petroleum generation and accumulation. Some of the areas of crustal instability became the loci of Tertiary deltas, and Klemme (1971) placed them in a separate basin category.

The intermontane basins of Canada all lie in the western Cordilleran region. Klemme (1971) distinguished two types of intermontane basins:

those transverse to the regional strike (IMT) and those parallel with
it (IMS). The Canadian basins are all parallel with the strike. These
deep narrow troughs have a characteristic irregular basal profile. Because of their provenance and high geothermal gradients, these basins
in Canada have a very low hydrocarbon potential.

Tertiary Megasequence

During the period of time represented by this megasequence, formation
of delta complexes was confined to UCM areas. The best, well-known example in Canada is the Mackenzie delta, on the northern coast. There are
several additional possible submarine-delta complexes off the mouths of
paleodrainage channels.

GIANT FIELDS IN MEGASEQUENCES

Although the system used seemed to be effective in classifying the
Canadian basins, the general applicability of this approach on a worldwide basis remained uncertain. We suspected from other work, such as
that of Klemme (1971), that megasequences would be meaningful map units
for continent-wide mapping and that the classification would identify
basin units of distinct character. A more general idea of the usefulness
and relative worldwide importance of the megasequences or time intervals
in identifying the events leading to basin and trap formation was obtained by relating the data on the giant oil and gas fields of the world
given by Halbouty et al (1970) to the classification scheme. The fields
were categorized according to basin types, trap-forming mechanisms, and
megasequences for purposes of statistical examination in terms of reserves
and number of fields. The conclusions that can be drawn from this comparison are tentative and directional because of the rather limited sample of the total number of fields, even though those used contain 81
percent of the oil and 60 percent of the gas reserves of the world. The
sample, however, included about 60 basins of all types, covering the
world; and, by considering the number of fields, one can avoid the heavy
weighting of the statistics by the supergiants. The data on the giant
fields appear to suggest that the same megasequences in many parts of
the world are characterized by the beginning of similar tectonic styles
involved in basin formation. Furthermore, the giant fields occur in
basins that fit into the classification scheme used for the Canadian
basins.

Figures 5 and 6 show that plots of the sizes of the world's giant
oil and gas fields, respectively, yield truncated, approximately lognormal distributions. These observations suggest that the giants and
supergiants are only part of a continuum, or the upper tail of a distribution, and are not in any way unique. The pie diagrams show the proportions of reserves of oil and gas by geologic system and by megasequence.
There is obviously a difference between distribution of oil and gas. The
bulk of the oil reserves (60.8 percent) is in the Jurassic-Cretaceous
megasequence. The Tertiary megasequence harbors an additional 29.6 percent, leaving less than 10 percent of the reserves of the giant oil fields
of the world in the other two megasequences. Similarily, the largest
proportion of gas (54.1 percent) is in the Jurassic-Cretaceous megasequence. However, although the Tertiary megasequence contains 24 percent
of the gas fields, only 11.5 percent of the gas reserves is in the Tertiary. On the other hand, 33 percent of the gas reserves and 30 percent
of the fields are contained within the upper Paleozoic - Triassic megase-

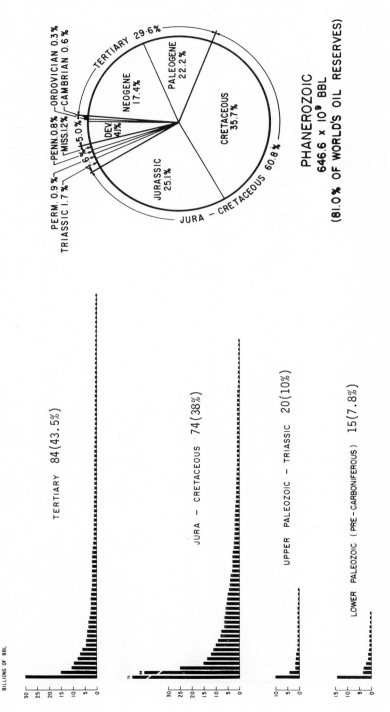

FIG. 5—Size distribution of giant oil fields (193) by megasequences. Number of fields is given after megasequence.

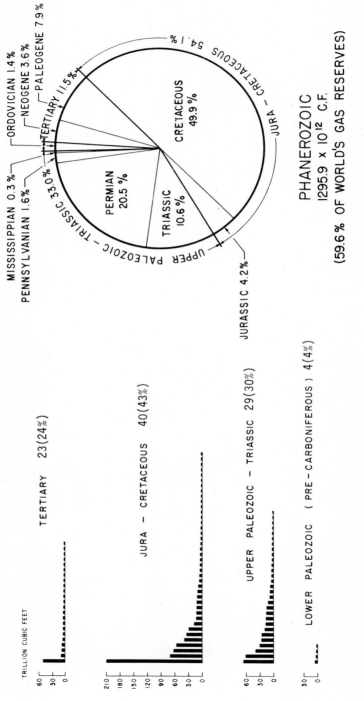

FIG. 6—Size distribution of giant gas fields (96) by megasequences. Number of fields is given after megasequence.

quence, compared to less than 3 and 10 percent, respectively, of the oil reserves and fields.

Certain other differences in the gas and oil data are apparent. First, there are only 96 gas fields in the giant category as opposed to 193 oil fields. Second, the gas/oil ratio between the giant-oil-field reserves and the giant-gas-field reserves is only about 2,000 cu ft/bbl. This ratio seems to be very anomalous compared to the gas/oil ratios for North American reserves, which are more in the order of 6,000 cu ft/bbl. It may mean that much gas has not been included in reserves estimates on a worldwide basis or may reflect the historically more vigorous search for oil. If one looks at the total proved recoverable oil compared to gas reserves for all fields of the world, the ratio is still only 2,800 cu ft/bbl.

Another observation perhaps of interest regarding the world's giant oil fields is the increasingly greater number of fields in the younger megasequences. There are 15 fields in the lower Paleozoic megasequence, 20 in the upper Paleozoic - Triassic, 74 in the Jurassic-Cretaceous, and 84 in the Tertiary. For gas, the picture is somewhat different; there are 4 giant gas pools in the lower Paleozoic, 29 in the upper Paleozoic - Triassic, 40 in the Jurassic-Cretaceous, and 23 in the Tertiary.

GIANT FIELDS BY BASIN TYPE

Figure 7 shows the reserves of the giant fields split according to the basin type within which they occur. The silhouette profiles are not at a common scale. Of the eight basin categories recognized, it is obvious that the CM basins are of overwhelming importance, for they contain 64 percent of the oil reserves and 83 percent of the gas reserves. The second largest category appears to be the structural modification of the fringe areas on the deformed side of this basin type (CMD). This type accounts for an additional 15 percent of the oil and 6 percent of the gas. The rift basins (R) assume the next order of importance—8 percent of the oil and 5 percent of the gas. Within the CC basin category, giant fields are almost nonexistent. These statistics are different from those of Klemme (1971) largely because of a difference in classification of a single case—the Arabian-Iranian basin. The dominance of the CM category results in part from a few basins which contain several supergiant fields. The number of giant oil fields in each basin category (not shown) reflects less dominance by the CM category (82 fields) and a significant number of fields in the CMD (34) and R (29) categories. SCM and CC basins still are insignificant for both oil and gas. A larger sample including at least the major fields would probably show the delta and intermontane basins to be of greater relative importance in number of fields.

The cursory examination of the giant fields of the world and the basins within which they occur suggests that the observations on the type and time of basin formation in Canada may have relevance on a much broader scale. Figures 8 and 9 show the distribution of reserves of the giant fields of the world by basin type for oil and gas, respectively, within each megasequence. Productive basin types are seen to be prominent in reserves in specific megasequences. For example, the basin styles related to the undeformed craton (CC, CM) appear throughout the Phanerozoic, whereas the rift basins or those of tensional origin (R) do not appear until the upper Paleozoic - Triassic megasequence. The coastal-margin (UCM, SCM) and intermontane basins (IMS, IMT) are introduced in the Jurassic-Cretaceous and become more important in the Tertiary.

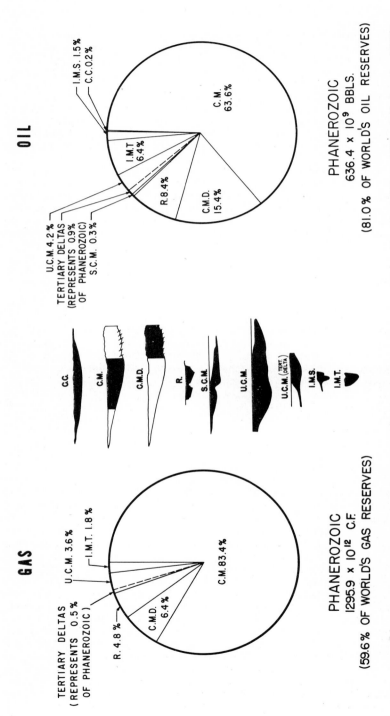

FIG. 7—Distribution of reserves in giant fields by basin classification. Basin examples shown: CC—Paris (modified after Héritier and Villemin [*in* Perrodon, 1972]); CM—Arabian-Iranian (modified after Hull and Warman, 1970); CMD—Arabian-Iranian; R—Gulf of Suez (modified after Heybroek, 1965); SCM—Blake Plateau (modified after Hayes and Ewing [*in* Perrodon, 1972]); UCM—Gulf Coast (modified after Lehner, 1969); UCM (Tertiary delta)—Niger (modified after Weber, 1971); IMS—Tofino; IMT—Maracaibo.

TRAP-FORMING MECHANISMS OF GIANT FIELDS

The trap-forming mechanisms have been divided into two broad categories—structural and stratigraphic. The structural category is divided
further into three main types—compressional (C), tensional (T), and
geostatic. The geostatic type includes, as subtypes, halokinesis (H),
lutokinesis (L), and gravity sliding (GS). The stratigraphic category
consists of two types—lithologic-nonexhumed reservoir (LITH) and sub-
unconformity or exhumed reservoir with reburial (SU). This fundamental
subdivision, which is purely genetic, should be helpful in relating tectonic evolution and basin formation to hydrocarbon accumulation. The
conventional classifications based on trap type tend to be descriptive
rather than genetic.

Figure 10 shows the giant-field reserves split according to the fundamental trap-forming mechanisms. The largest oil reserves are in traps
of the halokinesis subtype; tensional and compressional traps are also
important. Traps of the tensional type are overwhelmingly important for
gas. Stratigraphic-trapping mechanisms are relatively less significant
than the structural types; they account for 11 percent of the oil reserves
and 7 percent of the gas reserves.

The fundamental trap-forming mechanisms characteristic of the giant
fields apparently are related to certain basin styles and clearly are
associated with certain megasequences (Figs. 11, 12). It is also readily
apparent that there may be considerable differences between the mechanisms
related to gas and oil entrapment in the giant fields in a given megasequence. The data for the lower Paleozoic megasequence show the largest
proportion of oil and gas to be related to block faulting or tensional
structure. This observation should be further qualified. The time of
the faulting actually was during deposition of the subsequent upper
Paleozoic - Triassic megasequence and later. Though the rocks of the
trapping complex are of the lower Paleozoic megasequence, the actual
migration and trap filling occurred later. Other traps that might be
expected in the lower Paleozoic megasequence are the stratigraphic types
involving porosity pinchouts with or without unconformities.   •

In the upper Paleozoic - Triassic megasequence, there is a continuation of the trap types seen in the lower Paleozoic megasequence and a
dominance of tension-related structures. Halokinetic structures for both
oil and gas are first noted in this megasequence.

In the Jurassic-Cretaceous megasequence, 79 percent of the oil in the
giant fields is trapped by halokinetic structures. This dominance is
entirely a result of the large salt-induced structures of the Arabian-
Iranian basin. However, tensional structures are by far the most important for gas (95 percent), being even more important than in the upper
Paleozoic - Triassic megasequence because of tensional mechanics associated
with the West Siberian basins. Both the halokinetic and the block-faulted
structures occur within the craton-margin type of basin. Block faulting
is also important within the tensionally formed rift basins. Stratigraphic
traps, of course, are present throughout this megasequence, and the first
evidence of significant reserves associated with compressional movements
appears.

For the Tertiary megasequence, the most striking feature is the large
proportion of giant-oil-field reserves related to the compressional structures which are primarily in the foothills of the Zagros Mountains. In
addition, structures resulting from geostatic loading, such as lutokinetic
and gravity-sliding structures, are distinctive features of the UCM basins
including the Tertiary delta complexes.

The time range over which these various trapping mechanisms have been
operating to form the giant fields clearly varies. Figure 13 illustrates

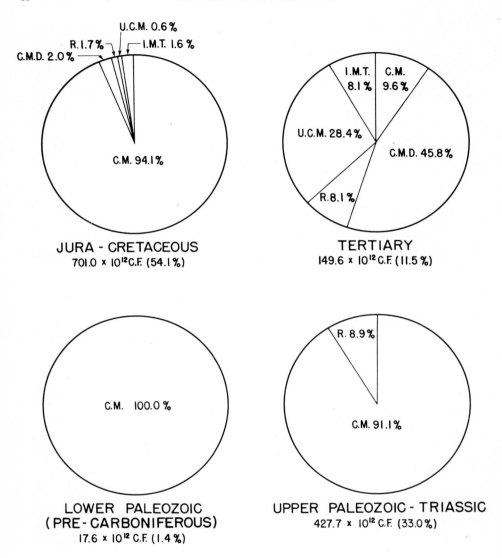

FIG. 8—Basin classification of giant-oil-field reserves by megasequences.

the chronology of trapping mechanisms. Throughout the Phanerozoic, there
was an increasing complexity of trap-forming mechanisms accruing with
time. This increment of types appears to be related to the major tectonic events dominating each megasequence. We suggest that an increasing
opportunity for entrapment as more types of trap-forming mechanisms
developed in the later megasequences may be an important factor in the
concentration of reserves in the Mesozoic and Cenozoic rocks.

   The structural trapping mechanisms suggest a contemporaneity between
age of reservoir and structural growth, whereas the stratigraphic trapping
mechanisms indicate that the geologic time interval between formation of
the host reservoir and emplacement of hydrocarbons may be great.

PROPOSED RESOURCE ESTIMATION SCHEME

We think that the system of basin examination developed for the Canadian study is applicable on a broader scale for the first stage in petroleum resource estimation, and we would make certain recommendations as a result. On the basis of our work, It appears that a considerably improved assessment of the world's petroleum resources would be possible if done as an international cooperative effort. An approach such as the following might be considered:

1. The first step should be the compilation of the regional geology within a broad time-stratigraphic framework such as the megasequences used herein.

2. On the basis of regional geology, the various basins should be defined and mapped. Those basins warranting more critical examination

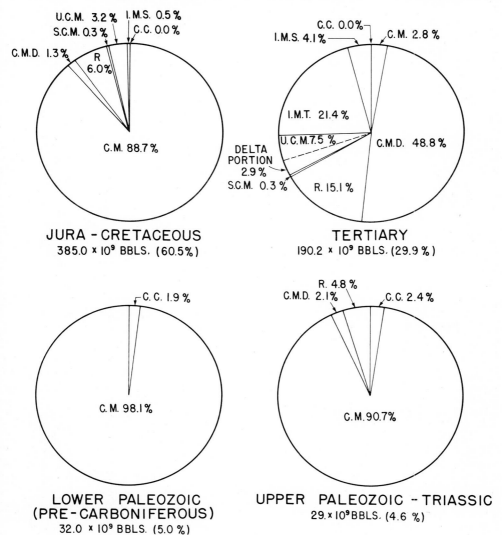

FIG. 9—Basin classification of giant-gas-field reserves by megasequences.

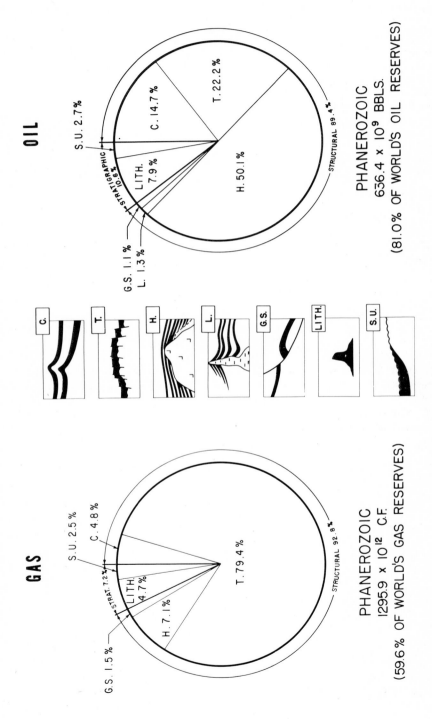

FIG. 10—Distribution of total Phanerozoic giant-field reserves by trapping mechanisms. Examples: compressional—Kirkuk (modified after Dunnington, 1958); tensional—Gronigen (modified after Stäuble and Milius, 1970); halokinesis—Bay Marchand (modified after Frey and Grimes, 1970); lutokinesis—West Barrackpore-Wilson (modified after Higgins, 1955); gravity slide—Western Gulf Coast (modified after Bruce, 1973); lithologic—Swan Hills (modified after Hemphill et al, 1970); sub-unconformity—East Texas (modified after Minor and Hanna, 1941).

should be determined. Implicit in this approach is the need for some general agreement on basin classification.

3. A qualitative rating should be made for the basins warranting more critical examination in terms of their petroleum potential. This rating should be made by use of a check list including all the important characteristics bearing on petroleum generation and accumulation that can be dealt with on a regional basis.

4. The next step would be a quantitative estimate of petroleum potential. For basins having very little or no exploration, this type of estimate would have to be made by analogy with other basins in the same family or category. The qualitative estimate from step 3 should be helpful in determining the specific placement of a particular basin within a range of yields for that basin type. This might best be expressed as a range in a probability curve. Volumetric-yield data should be much more meaningful when used in the context of a basin classification scheme. Basin yields based on proved reserves are of questionable value if not based on both oil and gas for basins in a mature stage of exploration, or appreciated in some way for immature basins to give the ultimate yield.

5. More detailed, definitive, and quantitatively meaningful estimates could be made for those basins having particular interest to the estimator within the broad geologic framework of the megasequences and a basin classification. Such estimates would involve a careful study of the basin geology in order to define the various oil and gas systems or "plays" within it. Where possible and desirable, the various parameters relating to petroleum accumulation could be described in quantitative terms (i.e., the characteristics bearing on the occupied reservoir volume under closure at a prospect level). These parameters can be postulated and based on actual or inferred data in the form of probability distributions and combined by a Monte Carlo simulation program into an ideal model which subsequently can be "risked" and cumulated into oil and gas systems and into estimates. In order for such an approach to be meaningful, however, all possible detailed information, including seismic sections, must be considered. This stage of estimating thus is not likely to be practical in the public realm, but can be done within a single organization.

CONCLUSION

In order to make rational estimates of potential petroleum resources, a regional geologic framework including a classification of the basins must be developed as a first stage in evaluation; and subsequent detailed studies must be consistent within this framework. The basin-analysis system developed for a study of the Canadian basins appears to be usable on a worldwide basis. A progressive increase of both preserved basin types and their associated trap-forming mechanisms seems to accrue with time. This increment of possible trapping situations is an important factor, in addition to preservation, not commonly recognized in the concentration of major reserves in the upper part of the geologic column.

REFERENCES CITED

Bruce, C. H., 1973, Pressured shale and related sediment deformation: mechanism for development of regional contemporaneous faults: AAPG Bull., v. 57, no. 5, p. 878-886.

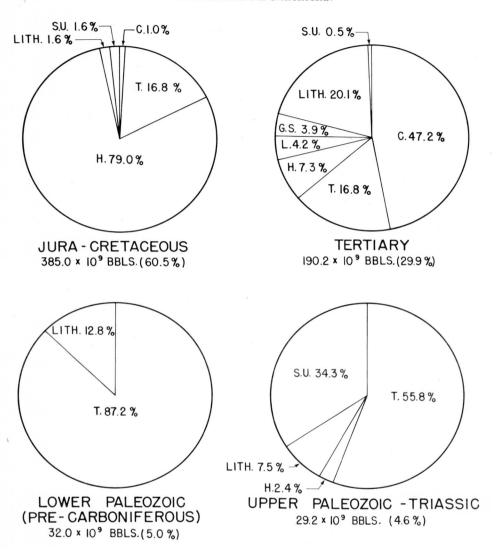

FIG. 11—Trapping mechanisms of giant oil fields by megasequences.

Dunnington, H. V., 1958, Generation, migration, accumulation, and dissipation of oil in northern Iraq, *in* Habitat of oil: AAPG, p. 1194-1251

Frey, M. G., and W. H. Grimes, 1970, Bay Marchand-Timbalier Bay-Caillou Island salt complex, Louisiana, *in* M. T. Halbouty, ed., Geology of giant petroleum fields: AAPG Mem. 14, p. 277-291.

Halbouty, M. T., et al, 1970, Factors affecting formation of giant oil and gas fields, and basin classification, Part II, *in* M. T. Halbouty, ed., Geology of giant petroleum fields: AAPG Mem. 14, p. 528-555.

Hemphill, C. R., R. I. Smith, and F. Szabo, 1970, Geology of Beaverhill Lake reefs, Swan Hills area, Alberta, *in* M. T. Halbouty, ed., Geology of giant petroleum fields: AAPG Mem. 14, p. 50-90.

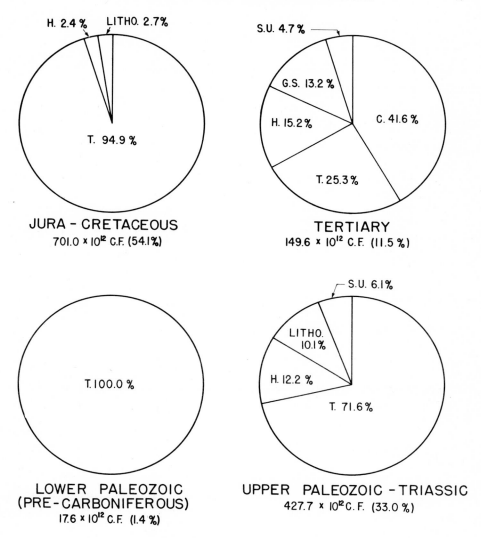

FIG. 12—Trapping mechanisms of giant gas fields by megasequences.

Heybroek, F., 1965, The Red Sea Miocene evaporite basin, *in* Salt basins around Africa: London, Inst. Petroleum, p. 17-40.

Higgins, G. E., 1955, The Barrackpore-Wilson oil field of Trinidad: London, Inst. Petroleum Jour., v. 41, no. 376, p. 125-147.

Hull, C. E., and H. R. Warman, 1970, Asmari oil fields of Iran, *in* M. T. Halbouty, ed., Geology of giant petroleum fields: AAPG Mem. 14, p. 428-437.

Klemme, H. D., 1971, The giants and the supergiants (3 parts): Oil and Gas Jour., March 1, p. 85-90; March 8, p. 103-110; March 15, p. 96-100.

Lehner, Peter, 1969, Salt tectonics and Pleistocene stratigraphy on continental slope of northern Gulf of Mexico: AAPG Bull., v. 53, no. 12, p. 2431-2479.

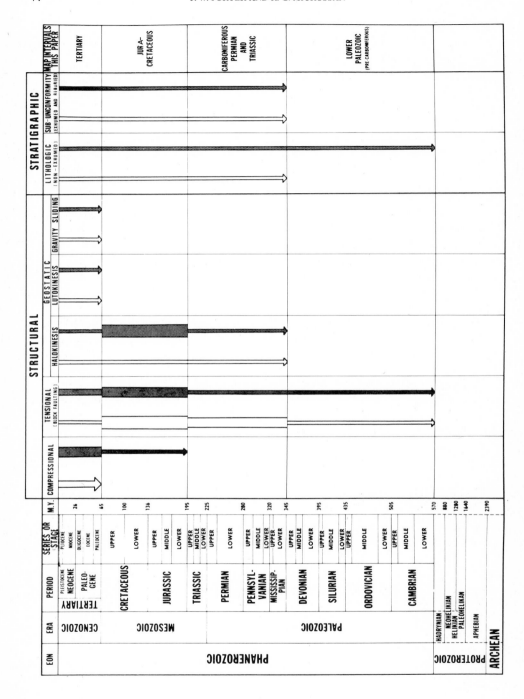

McCrossan, R. G., and J. W. Porter, 1973, The geology and petroleum potential of the Canadian sedimentary basins—a synthesis, *in* The future petroleum provinces of Canada—their geology and potential: Canadian Soc. Petroleum Geologists Mem. 1, p. 589-720.

Minor, H. E., and M. A. Hanna, 1933, East Texas oil field: AAPG Bull., v. 17, no. 7, p. 757-792.

Perrodon, Alain, 1972 (1971), Classification of sedimentary basins—an essay: Sci. Terre, v. 16, no. 2, p. 195-227.

Rainwater, E. H., 1972, Petroleum in deltaic sediments: 24th Internat. Geol. Cong. Trans., sec. 5, p. 105-113 (remark referred to in this paper was verbally).

Stäuble, A. J., and G. Milius, 1970, Geology of Groningen gas field, Netherlands, *in* M. T. Halbouty, ed., Geology of giant petroleum fields: AAPG Mem. 14, p. 359-369.

Weber, K. J., 1971, Sedimentological aspects of oil fields in the Niger delta: Geol. Mijnbouw, v. 50, no. 3, p. 559-576.

FIG. 13—Trap-forming mechanisms of giant fields classified by age of reservoir. Light arrows represent gas and dark arrows, oil. Width of arrows represents proportion of hydrocarbons in each category. Precambrian time scale is compressed 24 times Phanerozoic time scale.

Undiscovered Oil Reserves[1]

I. H. MACKAY[2] and F. K. NORTH[3]

ABSTRACT   That the traditional optimism of the oil industry can easily become irrational is perfectly exemplified by recent Canadian experience. Application of this experience to the world scene is illuminating.

Forecasts of "potential" reserves based on volumes of sedimentary rock, without the incorporation of probability factors, must be abandoned. It is impossible to quantify the unknowable by this form of analogy. The only type of forecast that can be based on experience takes into account the facts that (1) no readily accessible area of potential production remains in any real sense unexplored, and (2) the rate of addition to the Free World's conventional reserves appears to be established on a negative slope.

The production-depletion curve can be extended beyond its mathematically determinable "tail" only by the discovery of basins capable of very prolific production (e.g., basins such as the North Sea basin). The type of oil field characteristic of North America now makes no perceptible difference in the projections. Such new productive basins almost inevitably will be in remote, hazardous, or insecure regions, largely offshore and in deeper or more dangerous waters. Operation in such areas requires not merely large fields, but fields with thick "pay" zones, so that fixed platforms can be used. Such fields are almost bound to be in structural traps; in offshore areas, structural traps are very likely to be diapiric and should be identifiable early in the exploration of the basins.

Geographic, political, engineering, and other factors determine the producibility characteristics required to make any new basin economically viable compared with alternative sources of hydrocarbons (comparable production from oil ["tar"] sands, oil shales, or coal). Empirically derived probability factors then must be incorporated into the quantities of production required to extend the tail of the production-depletion curve. These probability factors appear to be of the following order: (a) probability that an explored basin will yield a single field of 500 million bbl, about 1 in 6; (b) probability that an explored basin will yield a single field of 1 billion bbl, about 1 in 10; (c) probability that an explored basin will yield one 1-billion bbl field capable of development by about 64 wells (4 times 16 well platforms), about 1 in 20 or less; (d) probability that an explored basin will contain more than 5 billion bbl of recoverable reserves (to justify a 30-in. pipeline with 20-year life), about 1 in 14; (e) probability that an explored basin will contain more than one supergiant oil field, about 1 in 45.

With risk factors of these orders incorporated into projections, the odds may now be that the Free World's undiscovered reserves of conventional oil are smaller than the amount already produced.

INTRODUCTION

The writers do not believe that detailed geologic controls of oil or gas occurrence can be discussed usefully in the context of estimating undiscovered reserves. Whether traps were formed early enough, or whether caprock or permeability is adequate, can really be revealed only by drilling, and *all* accessible sedimentary areas will be drilled. All experience shows that highly petroliferous basins are consequences either of crustal compression or crustal extension; they are spatially and temporally associated either with mountain building or with continental drift. To say that "plate tectonics" can help to discover more oil or gas is, however, essentially teleological. Plate tectonics certainly cannot help us in estimating the volumes of undiscovered oil or gas. We cannot quantify the unknowable by any application of an inexact science.

Far more significant in this context is experience of the actual record. The record in the United States has been clear for nearly 20 years, and that in Canada has been clear for 5 years; that for the world as a whole (or at least for the Free World) is becoming equally clear.

[1] Manuscript received, January 27, 1975.
[2] Oil and Gas Department, Bank of Montreal, Calgary, Alberta.
[3] Department of Geology, Carleton University, Ottawa, Ontario.

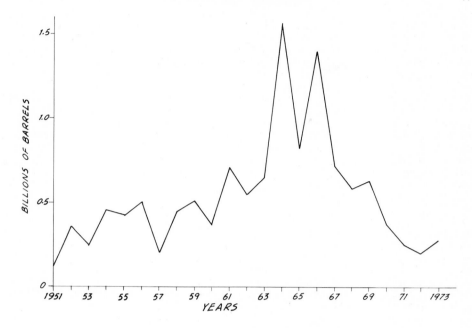

FIG. 1—Total annual additions to Canadian oil reserves.

The greatest single obstacle to an appropriate reaction to this knowledge
has been the ingrown habit of blaming various levels of government for
a great industry's failure to understand the foundation of its own business.

METHODS OF ESTIMATING RESERVES

Methods of estimating ultimate reserves of any primary resource (other
than methods that are wholly theoretical) seem to fall into four categories:

1. Use of average crustal concentrations of the resource (McKelvey,
1972). However, oil and gas, unlike iron formations or porphyry coppers,
are discretely distinct from the rocks that contain them. Moreover,
recovery constraints are critical; Gonzalez's (personal commun. to K. H.
Crandall, January 17, 1974) point about gold in seawater is well taken:
"Trace elements of gold in the waters of the oceans mean a vast existence
of known gold—more than a million dollars per capita for all the inhabi-
tants of the earth—but no one has figured out how to recover that gold
at a cost less than its value, and if they did the value would change."

2. Use of volumetric analogies (variations on the "Weeks method").
These methods require the incorporation of probability factors ("risk
factors"), but they cannot themselves indicate what those factors should
be.

3. Use of rate analogies according to *input* (effort), such as the
Zapp method (Zapp, 1962). Such methods have been decisively discredited.

4. Use of rate analogies according to *output* (results). The Hubbert
method (Hubbert, 1966) belongs here.

The last method, as a general technique, has been derided but has
not so far been discredited. Indeed, the lesson it held for the United

States was learned about 12 years too late (or, by some, not at all).
At the time of its first application, the American oil industry was in
the midst of the most extraordinary frenzy of drilling in its extraordinary history. In each of the 15 years from 1950 to 1964, more than 40,000
wells were drilled in the lower 48 states. Nearly 200,000 of the total
wells were classed as exploratory; they discovered just one field big
enough to be operable in any "frontier" region (West Delta Block 73,
Louisiana). At the midpoint of this drilling splurge, the oil-import
quotas were imposed. Their tacit removal, at the time of Interior
Secretary Hickel's delegation to Ottawa at the end of 1970, represented
the first clear acceptance of a conclusion that should have been accepted
a decade earlier. Precisely this same outcome is now apparent for Canada,
still regarded by many people as an inexhaustible resource base. Furthermore, the same exhaustibility of petroleum resources is now discernible
for the world as a whole.

ESTIMATES OF FUTURE PETROLEUM RESERVES

Figure 1 shows the historic rate of addition to oil reserves in
Canada. The curve assumed a negative slope about 1964. This permitted
the forecast that Canada's recoverable reserves would start to decline
by 1970. When this forecast proved true, it became possible to predict
that Canada's production would start to decline not later than 1976.
Almost everyone now acknowledges this eventuality, in spite of the con-

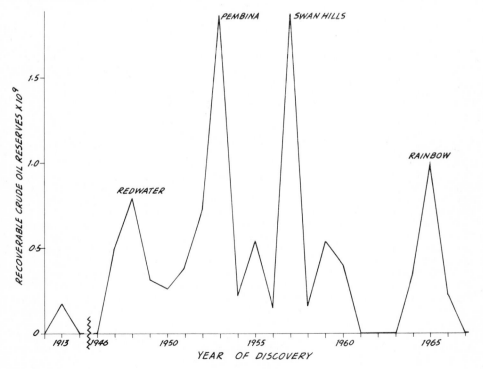

FIG. 2—Discovery and reserves of major oil fields (more than 100 million bbl recoverable) in Canada.

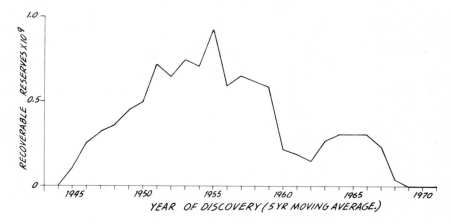

FIG. 3—Reserves of Canadian major oil fields credited to year of discovery; 5-year moving average.

tinued claim that about 100 billion bbl of reserves still can be found and produced.

Figures 2 and 3 show the pattern of discovery in Canada for major fields only; reserves are assigned to the year of their field's original discovery. Figure 4 shows the equivalent pattern, for Alberta only, according to drilling activity. We must hope that our countrymen will not be as slow to recognize the obvious trend of declining output as the Americans were.

Figure 5 shows the equivalent pattern of development for the oil fields of the whole world. Figure 6 shows reserves of fields larger than 100 million bbl of recoverable oil. Figure 7 shows 5-year moving averages of world oil consumption for the different classes of all major oil fields. From these data, the following observations seem justified.

1. Since 1948, the average rate of oil finding in the Free World has been about 18 billion bbl annually. Although there are inevitable fluctuations, the rate has shown a perceptible decline since 1963. The continued increase in remaining reserves is due far more to extensions and revisions of known fields than to new discoveries.

2. The *average* size of fields discovered has declined since about 1953. Though the largest single oil fields in North America, Europe, the Soviet Union, Africa, and Australia were discovered during the past 10 years, there has been an unmistakable decline in the discovery of supergiant fields. Of the approximately 100 oil fields having recoverable reserves in excess of 1 billion bbl, nearly half (41) were discovered in the single decade 1955-1964, aptly categorized by Warman (1973) as the "heyday" of exploration in the Middle East and North Africa.

3. Outside the eastern-bloc countries, less than 5 percent of remaining proved oil reserves is in *basins* found to be productive since the end of the 1950s.

4. The demand curve (now at 20 billion bbl per year; Fig. 7) overtook the 5-year-moving-average curve of discovery for large fields in 1967—for the first time in this century. The reserves/production ratio is now declining dramatically. For conventional petroleum the ratio is now below 30. If the historical rate of increase in demand of 7.5 percent is continued, the reserves/production ratio for the Free World, generously defined, is less than 15 and approaching the "peril point."

5.   This rate of increase in demand cannot be sustained from known reserves beyond the end of the 1970s. Oil production actually will decline by about 1985, when the world will be nearly 20 million bbl/day short if present rates of discovery and growth continue.

1.   The world demand is now such that only large fields, in prolific basins, are of any significance. North American domestic experience, derived from tens of thousands of very small fields, is a most treacherous guide in these new circumstances.

2.   Large fields and prolific basins are sufficiently few in number, with respect to the "average," so that statistical generalizations are of some validity. They provide empirical *probability factors* for various expectations of producibility.

It is our belief that the present rate of discovery is in fact likely to decline still further, rather than to resume a positive slope. Areas of potential new supply are being reduced even more rapidly than demands are increasing. This circumstance, which applies to all resource development, is exacerbated when the resource concerned is finite and nonrenewable. It is now imperative that we adopt policies that acknowledge three inescapable constraints:

3.   No readily accessible region is now unexplored. Large fields in prolifically productive basins are now apt to be found only in remote, hazardous, unreliable, or otherwise "sensitive" regions. These factors raise formidable questions of what is economically and technologically sensible. The *probability factor* is thus overlain by a *feasibility factor*—the sum total of the political, economic, and engineering constraints that tend to offset the *net energy gain* from the operation.

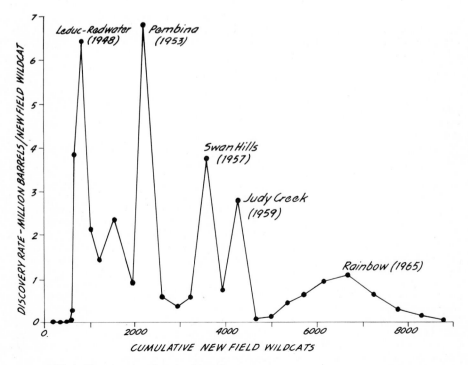

FIG. 4—Discovery rate of crude oil in Alberta, per new-field wildcat (from Ryan, 1973).

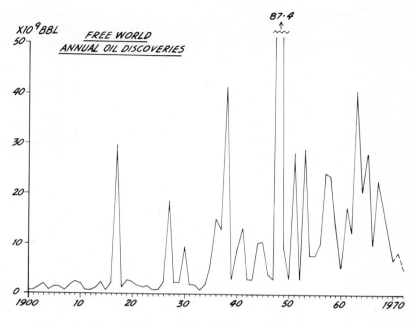

FIG. 5—Free-world annual oil discoveries (prepared by H. R. Warman, 1971, 1973).

Large Fields

A field that is areally large but has a thin "pay" zone (e.g., East Texas or La Brea - Parinas) requires far too many development wells to be economically viable in the Canadian Arctic or in a remote offshore location. In waters deeper than about 100 m, fields must be developed from a small number of fixed platforms; thus they must have thick "pay" zones. As Klemme (1973, 1974) has emphasized, such "pay" zones imply structural traps. Because continental shelves, almost by definition, are undeformed orogenically, large structural traps are very likely to be diapiric. They should be indicated by the first seismic lines that cross them. Exploration for "subtle stratigraphic traps" after all the big structures have proved dry (e.g., on the Grand Banks of Newfoundland) is unlikely to provide improvement to the overall supply picture.

Feasibility Factors

A remote oil or gas field needs a delivery system. Long pipelines are highly inflexible and very expensive. Their sizes are dictated by the reserves discovered; the sizes in turn dictate the volumes that can be transported.

Suppose that a field the size of Leduc or Kelly-Snyder were discovered in the middle of the Beaufort Sea 12 months from today. What would happen then? How long would we wait, and how many more wells would we drill (against the opposition of all environmentalists and a great many other influential people), before we decided how many platforms to erect? How large a pipeline should be built? During building of the pipeline, suppose another discovery were made, this time 50 mi (80 km) off the coast of Labrador. Would we start to build a second pipeline? How much should we spend; how many rigs should we move to the two areas; how large and expensive a delivery system could we construct, *before the net energy*

turnover would become negative? Where is the point of no return for
economic feasibility?

The industry is well aware of this, you say? Canada's National Energy
Board just finished (May 1974) a series of hearings concerning the export
of Canadian oil to the United States. Every large oil company operating
in Canada made a submission. Nine of these industry submissions included
graphic forecasts of oil production in Canada to 1990 or beyond. As
these forecasts had to be reconciled with the one submitted by the Canadian Petroleum Association, it is not surprising that they were all very
much alike. They are physically, economically, geographically, and in
all other respects impossible of achievement—even if the frontier oil
they envisage exists to be discovered, which is doubtful. It remains to
be determined what the feasibility factor must become before the decision
is made to turn to alternative sources.

Probability Factors

The feasibility factor dictates the minimum size of field viable
within any new basin and the minimum volume of recoverable reserves for
the basin itself. In all areas of future exploitation, both basinal
yields and individual field sizes will have to be large.

Only about 40 decipherably distinct basins in the world are known to
contain even a single oil field larger than 500 million bbl (recoverable
oil). The empirical probability that an explored basin will yield even
one such field is therefore about 1 in 6 or 7. The empirical probability
of finding a basin with more than one such field is less than 1 in 10.
This explains the sixfold to tenfold overestimation of the "potential"
reserves of Brazil, India, and Australia (and Canada!), resulting from
the algebraic summation of volumetrically estimated reserves for the considerable number of discrete basins within each of these territories.
In addition, the probability that an explored basin will yield a single
field of 1 billion bbl (recoverable oil) capable of development by about
64 wells (4 times 16 well platforms) is about 1 in 20 or less.

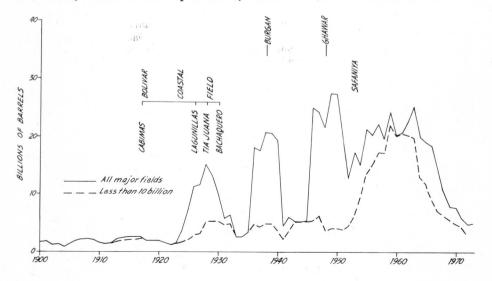

FIG. 6—Reserves of major oil fields of the world (more than 100 million bbl recoverable), credited to
year of discovery; 5-year moving average.

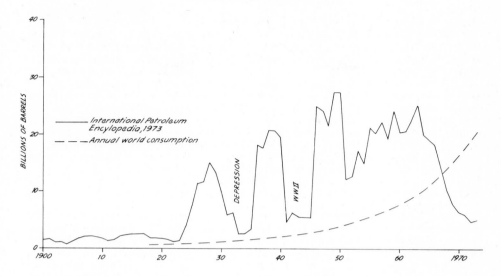

FIG. 7—World oil consumption versus discovery of major fields, credited to year of discovery; 5-year moving average.

If, for geographic or other reasons, a basin could not be economically viable without a production rate of 500,000 bbl/day through a long, 30-in. pipeline, the empirical probability that it will be prolific enough to achieve this status is about 1 in 12. Figure 8 shows graphically the empirically derived probability factors for different levels of producibility.

More than half of the volume of recoverable oil already discovered is in supergiant fields; the probability that an explored basin will contain more than one supergiant field is about 1 in 45.

Future Outlook

Klemme (1971) estimated that there are about 118 undeveloped basins in the world, lying offshore or beneath inland seas. We doubt whether all of these are true sedimentary basins which contain abundant marine strata and which are practicably accessible to the drill in the near future. However, even if all of them are explored in the balance of this century, the factors just enumerated indicate that it would be fortuitous if more than about 10 of them were productive enough to make any impact on supply. If these 10 basins all were as good as the Indonesian or Alberta basin, we might expect a yield of new oil between 100 and 150 billion bbl—an extension of the life expectancy for world oil of about 6 years.

So far, about 850 billion bbl of recoverable, conventional oil has been discovered. More than half of this volume is in only 33 supergiant fields, each with more than 5 billion bbl of recoverable oil. Estimates of ultimate recoverable crude oil as high as 2,000-2,500 billion bbl (Hendricks, 1968; Weeks, 1971) therefore imply about 1,200 billion bbl still to be discovered.

Not only do we think it highly unlikely that more conventional oil remains to be discovered than has been discovered so far, we even think it unlikely that there is much more remaining to be discovered *and exploited* than has been *consumed* so far. If only large fields in prolific basins are capable now of making any impact, the reserves volumetrically estimated to remain for future discovery are probably overstated by a factor of at least six. They are more likely, therefore, to be of the order of 150-200 billion bbl than the 800-1,200 billion bbl popularly estimated.

If the "Hubbert pimple" is valid as a representation of the life pattern of a finite resource, the bell curve reached its midpoint about 1960. If the industry does as well after 1960 as before 1960 (a nearly unimaginable circumstance), and if we exclude for the moment the rare, supergiant fields having more than 5 billion bbl of recoverable hydro-carbons each, then about 220 billion bbl remained to be discovered after 1960. As more than 190 billion bbl had been discovered between 1960 and the end of 1973, only about 30 billion bbl may remain to be found in fields exclusive of supergiants. Even if we assume that each of the 10 potentially prolific, "new" basins will contain a "Prudhoe Bay," we have reserves totaling less than 150 billion bbl.

Hunt (1962), Wickman (1956), Conybeare (1965), and many others have made highly speculative calculations of ultimate oil resources based on the volume of organic carbon in the world's sedimentary rocks. The commonest figure for ultimately recoverable oil (excluding gas) to re-sult from these calculations is 3,000 billion bbl. At least half of this volume is thought to be contained in heavy-oil sands or present as surface seepages, so the figure for recoverable conventional oil, although con-

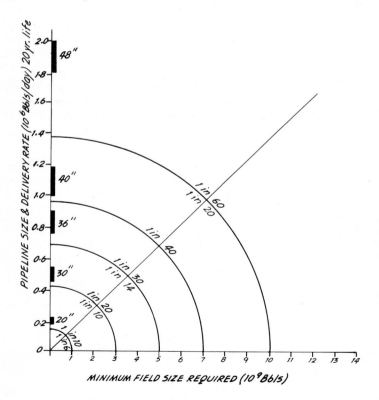

FIG. 8—Chance of new basin yielding a single field of stated size, with equivalent pipeline requirements. Top number is probability of explored basin yielding a single field of given size. Bottom number is proba-bility of explored basin having a total reserve of given size.

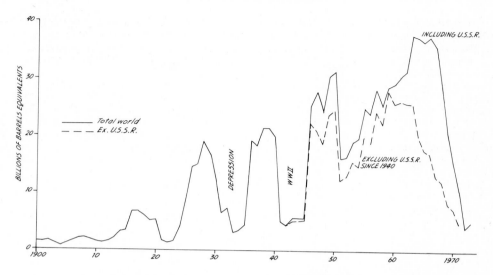

FIG. 9—Major oil and gas fields of world, credited to year of discovery; 5-year moving average, with and without Soviet Union. Gas in oil-equivalent barrels.

siderably higher than ours, is nonetheless closer to it than to the higher estimates.

NATURAL GAS RESERVES

We regard estimates of world ultimate reserves of natural gas as little more than guesswork. Studies based on past experience are far less adequate for gas than for oil. Enormous quantities of natural gas found in association with early oil fields were flared. We have no real records of the amount produced (and wasted) before the late 1940s. We know that natural gas is much more widespread, both geologically and geographically, than is conventional oil, but we do not know what conditions are favorable for very large accumulations of nonassociated gas (like those in Siberia). Discussion of possibly favorable conditions has already been made elsewhere by one of us (North, 1972) and it will not be recapitulated here.

The standard ratio of 6,000 or 6,500 cu ft of gas per barrel of oil has been based on recorded cumulative *production*, which was achieved under the distorted conditions just described. As though all important gas reserves were somehow ancillary to oil fields, some authorities (apparently including the U.S. Geological Survey) now calculate *gas in place* from *oil in place*. Because the relation between these two quantities has not been adequately demonstrated, and because oil in place is an unknown volume of uncertain meaning, this calculation is almost certainly meaningless. Just as gas reserves must now be sought independently of oil reserves, so forecasts for the two fuels must be made independently (when that becomes possible).

Figure 9 shows the pattern of discovery of all hydrocarbons, with gas converted to oil equivalent at 6,000 cu ft/bbl. If the Soviet Union is excluded, the pattern is one of precipitous decline since 1959. Even including the great gas discoveries in western Siberia in the 1960s, the worldwide pattern provides a declining curve since 1966. Although natural gas was the last of the fossil fuels to be widely used, it seems inevitable that it will be the first to be depleted.

REFERENCES CITED

Conybeare, C. E. B., 1965, Hydrocarbon-generation potential and hydro-
    carbon-yield capacity of sedimentary basins: Bull. Canadian Petro-
    leum Geology, v. 13, p. 509-528.
Hendricks, T. A., 1965, Resources of oil, gas, and natural-gas liquids
    in the United States and the world: U.S. Geol. Survey Circ. 522, 20 p.
Hubbert, M. K., 1966, History of petroleum geology and its bearing upon
    present and future exploration: AAPG Bull., v. 50, p. 2504-2518.
Hunt, J. M., 1962, Geochemical data on organic matter in sediments: Proc.
    Internat. Scientific Oil Conference, Budapest.
International Petroleum Encyclopedia, 1973: Tulsa, Petroleum Publishing
    Co.
Klemme, H. D., 1971, Giants, supergiants, and their relation to basin
    types: Oil and Gas Jour., March 8, p. 103-110.
——1973-1974, Structure-related traps expected to dominate world-reserve
    statistics: Oil and Gas Jour., December 31, 1973, and January 7, 1974.
McKelvey, V. E., 1972, Mineral resource estimates and public policy:
    Am. Scientist, v. 60, p. 32-40.
North, F. K., 1972, A sane look at U.S. gas resources: Natl. Gas Survey,
    U.S. Federal Power Commission, v. 5, 1973, p. 113-156.
Ryan, J. T., 1973, An analysis of crude-oil discovery rate in Alberta:
    Bull. Canadian Petroleum Geology, v. 21, p. 219-235.
Warman, H. R., 1971, Why explore for oil and where?: APEA Jour., v. 11,
    p. 9-13.
——1973, The future availability of oil: Conf. on World Energy Supplies,
    Grosvenor House, London, September 18-20: 11 p.
Weeks, L. G., 1971, Marine geology and petroleum resources: 8th World
    Petroleum Cong. Proc., v. 2, p. 99-106.
Wickman, F. E., 1956, The cycle of carbon and the stable carbon isotopes:
    Geochim. et Cosmochim. Acta, v. 9, p. 136-153.
Zapp, A. D., 1962, Future petroleum producing capacity of the United
    States: U.S. Geol. Survey Bull. 1142-H, 36 p.

Petroleum-Zone Concept and the Similarity Analysis— Contribution to Resource Appraisal[1]

CHRISTIAN BOIS[2]

ABSTRACT   A *petroleum zone* is a sedimentary volume which contains pools showing several common characteristics. Evaluating the potential of any area involves the forecasting of *potential petroleum zones* in the area on the basis of geologic parameters. Such potential petroleum zones can be evaluated by comparison with numerous producing petroleum zones in the world, through similarity computation and cluster analysis.

INTRODUCTION

Analogy is a method widely used for resource appraisal and generally involves large sedimentary units such as basins or provinces. Because these units may contain various types of hydrocarbon habitats, the analogy between two large units may be difficult to determine. At best, a resource appraisal can provide no more than an average estimate of the total resources of an area. Important parameters such as distribution of pool sizes, recovery factor, flow rate of wells, and chemical composition of oils cannot be assessed properly because of the large variations in these parameters from one part of a unit to another. Use of a smaller sedimentary unit should improve the petroleum evaluation, and the concept of a petroleum zone has been devised as a basis for evaluations.

PETROLEUM ZONES

A petroleum zone is defined as a continuous portion of sedimentary volume which contains pools showing the following common characteristics: (1) reservoirs within the same productive sequence occur throughout the zone; (2) hydrocarbons are of similar chemical composition; and (3) traps are of the same type, or belong to a small number of types. According to this definition, a petroleum zone is limited in extension both geographically and stratigraphically, and may be considered approximately equivalent to a "play" (Fig. 1). Accordingly, two petroleum zones can be partly or entirely superposed.

Groups of similar pools which form petroleum zones may be recognized in several important oil- and gas-producing areas—e.g., the Middle East, Venezuela, or the Permian basin. A continuously expanding file, being prepared by geologists in France, presently contains data on more than 80 petroleum zones and is aimed at covering many types of hydrocarbon habitat in the world.

The petroleum zones are described by approximately 100 parameters. Half of them concern the geologic situation—age of the petroleum zone, thickness of the series, stratigraphy, tectonics, type of reservoirs, source rocks, etc. The rest of the parameters describe the hydrocarbon distribution—reserves, number of pools and fields, producing acreage,

[1]Manuscript received, January 20, 1975.
[2]Institut Francais du Pétrole, 92502 Rueil Malmaison, France.
This paper summarizes the work carried out by a team of geologists in Institut Francais du Pétrole and ELF and TOTAL oil companies. A more extensive paper on this subject will be presented before the 1975 World Petroleum Congress.
The writer is very grateful to R. G. McCrossan and N. L. Ball, Geological Survey of Canada, Calgary, for their assistance and helpful suggestions after reading the manuscript.

number and thickness of reservoirs, flow rate of wells, type of hydrocarbons, etc. The non-numerical parameters have been transformed into
numerical codes, so that all parameters can be processed by a computer.

SIMILARITY COMPUTATION

   In a particular area, one or several *potential petroleum zones* may
be postulated and delineated according to the definition. These zones,
which are described by geologic parameters only, may be compared with
petroleum zones on file by means of a similarity computation.
   The similarity between two zones is maximum (100 percent) where every
parameter has the same value for both zones. Where the difference between
the values of both zones is maximum for every parameter, the similarity
is minimum (0 percent). The similarities are a function of the actual
differences between the couple of values shown by each parameter. A
list of the computed similarities is used in order to facilitate complete
identification of the similarities between the potential petroleum zone
and the zones of the file.
   Actually, the result of the computation depends on the description
of the parameters, the transformation and the weighting of the primary
descriptions, and the mathematical processing of the data. In order to
check the practical value of the method, similarities between every pair
of zones (among 60 zones from the file) have been computed using four
different similarity coefficients and various transformations and weight-
ings of the geologic parameters. The results of these different computations have been studied by cluster analysis (complete linkage) in order
to compare the results of the different computations. This analysis
(Fig. 2) put the zones having strong similarities into the same cluster.
For example, petroleum zones 9-37 fall within cluster C. The level of
similarity is shown by the horizontal lines which link the different
petroleum zones of the cluster. The minimum level of similarity for all
the petroleum zones in cluster C is 66 percent, as indicated by the lowest
common horizontal line linking the cluster. If a petroleum zone from

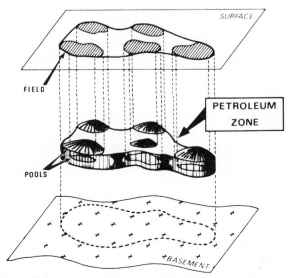

FIG. 1—Relations among hydrocarbon pools, fields, and petroleum zones.

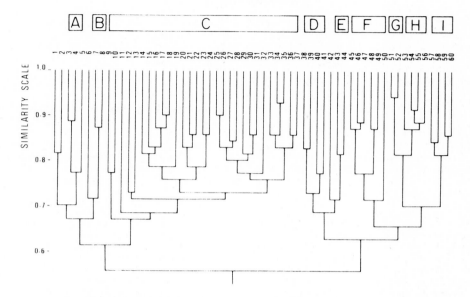

FIG. 2—Cluster analysis of similarities among 60 petroleum zones, based on geologic parameters only. Letters A to I are nine clusters which persist when different computation procedures are used.

cluster C is compared with a zone from any other cluster, the similarity will be less than this minimum value.

The results of different types of computations demonstrate that the method is valid for comparing petroleum zones. Although the graphs corresponding to the different computations differ, 90 percent of the zones are always distributed in the same nine clusters. These clusters are representative of typical geologic situations. Indeed, several geologic parameters within a given cluster have distributions with a much narrower range than that of the complete file. Several hydrocarbon parameters also show the same situation. Cross correlations and multiple correlations between geologic and hydrocarbon parameters within the largest clusters presently are being studied. The relations between the two types of parameters will enable predictions to be made on possible hydrocarbon characteristics in a petroleum zone for which only the geologic parameters have been defined.

CONCLUSION

To evaluate the hydrocarbon possibility of an area, all potential petroleum zones must be considered. After their delineation, such potential petroleum zones may be compared with every zone in the file, by using known or assumed parameters. The greatest similarities computed identify the four or five zones of the file which are most similar to the potential petroleum zone and provide examples of hydrocarbon habitats that are possible in the studied area. In addition, if a potential petroleum zone falls into one of the clusters of the file, the characteristics of this cluster may be used for more precise prediction.

By taking into account many parameters, the method permits the objective selection of examples which may guide the geologist in appraisal of hydrocarbon resources.

Potential Gas Committee and Undiscovered Supplies of Natural Gas in United States [1]

B. W. BEEBE,[2] R. J. MURDY,[3] and E. A. RASSINIER[4]

ABSTRACT   The mission of the Potential Gas Committee (PGC) is the monitoring of undiscovered supplies of natural gas in the United States and periodic publication of its estimates. *Potential supply* of natural gas is that quantity of natural gas yet to be proved (as the term *proved* is used by the American Gas Association's Committee on Natural Gas Reserves) by wells to be drilled under conditions of adequate but reasonable prices and normal improvements in technology.

Estimates of potential supply are made by members of 11 Work Committees for 12 areas which may include one or more geologic provinces. The results are published periodically for the guidance of government, the gas industry, and financial institutions. The guidelines for preparing estimates are flexible, and may change with improved technology and greater incentive. For example, undiscovered supplies of natural gas such as those locked in "tight" sandstones in several basins in the Rocky Mountain region will be included in the estimates when technologic and economic feasibility is demonstrated. At this time, drilling depths are limited to 30,000 ft (9,144 m) and water depth to no more than 1,500 ft (457 m).

The estimates are defined by decreasing degrees of certainty: "probable," "possible," and "speculative." The probable supplies are closely related to proved reserves and result from extensions to known gas deposits and new-pool discoveries. Possible supplies result from new-field discoveries in formations previously productive in a particular geologic province. Speculative supplies, which are the least certain, may lie in formations not previously productive in a productive province or in a province in which there is no production.

Undiscovered supplies do not include proved reserves estimated by the AGA Committee on Natural Gas Reserves. PGC and the AGA Committee are completely separate entities, but communicate through their respective chairmen to avoid duplications.

The PGC uses the geologic-volumetric or "attribution" technique developed by Lewis G. Weeks and others for estimating undiscovered oil, but some refinements have been made in the technique. Estimates are not based on the gross rock volume of each province, but on volume of individual zones producing, condemned, and potential. Estimates of speculative supply are based on character of potential producing sediments and by comparison with productive zones or provinces with similar characteristics. Estimates are checked against historical and statistical data and other pertinent information available to the Work Committees. Obviously, judgment plays an important role in formulating estimates of undiscovered gas supplies. Each member of a Work Committee has a high degree of expertise in the local area to which he is assigned. Workshops for members of one or more Work Committees are held periodically to develop uniformity in estimating techniques and to evaluate continuously the judgment factor. A Special Projects Committee undertakes assignments which may be beneficial to the PGC in making and evaluating its estimates. An Editorial Committee organizes the report for publication.

The PGC consists of over 100 individuals, mostly geologists and engineers, from all branches of the gas industry including distribution, production, and pipelines, both interstate and intrastate. In addition, there are observers from the USGS, FPC, EPA, NARUC, AGA, INGA, API, and the Office of Oil and Gas. Important changes are made as a result of consensus, and not by simple majority, and then only after spirited discussion and careful consideration. The Committee is a close-knit, dedicated group of outspoken, rugged individualists.

Despite the periodic publication of a wealth of superior, detailed information which is unavailable from any other source, the PGC is not without critics. Perhaps the criticism most commonly made is that the Committee makes no estimate of the rate at which its estimated undiscovered supplies of natural gas can be converted to reserves.

The Committee is currently reviewing its methods, format, frequency of publication, possible additions and deletions to its reports, and other aspects of its work to make its efforts more productive and useful.

[1] Manuscript received, January 31, 1975. Text reprinted, with minor changes, from Potential Gas Committee Report, "Potential Supply of Natural Gas in United States (as of December 31, 1972)": November 1973, Colorado School of Mines Foundation, Inc., Golden, Colorado.

[2] Consultant, Boulder, Colorado 80302.

[3] CNG Producing Company, Clarksburg, West Virginia 26301.

[4] Trunkline Gas Company, Houston, Texas 77001.

GUIDELINES FOR ESTIMATION OF POTENTIAL SUPPLY
OF NATURAL GAS IN THE UNITED STATES

General Policy Statement

An appraisal of the long-range prospects of the natural gas industry in the United States—an appraisal that is useful for financial, managerial, and government purposes—requires scientific, authoritative, and objective estimation of the potential supply of natural gas which may become available to the industry, in addition to proved recoverable reserves of natural gas currently available. The Committee on Natural Gas Reserves of the American Gas Association, an industry committee, is fulfilling part of this requirement by providing periodic estimates of proved recoverable reserves of natural gas immediately available to the industry.

The Potential Gas Committee, an industry committee supervised by the Mineral Resources Institute of the Colorado School of Mines Foundation, Inc., will fulfill the remainder of this requirement by providing periodic estimates of the potential supply of natural gas which may become available to the industry. This Committee, together with its several Work Committees, is composed of more than 100 individuals from the production, transmission, and distribution segments of the natural gas industry who are knowledgeable with respect to the potential supply of natural gas.

The Potential Gas Committee and its Work Committees discuss matters and data that are confidential. Fulfillment of the objectives of the Potential Gas Committee cannot be realized without the availability and utilization of information of a highly confidential nature. It is, therefore, the policy of the Potential Gas Committee and its Work Committees that every member, observer, and representative will respect such information and that there will be no disclosure of information or estimates, discussions, or other data, except as approved for release by the Potential Gas Committee.

DEFINITIONS AND CONCEPTS OF POTENTIAL SUPPLY OF NATURAL GAS

Definition of Natural Gas

Natural gas, as used in this report, is any gas of natural origin that is composed primarily of hydrocarbon molecules producible from a borehole. Most natural gas contains some nonhydrocarbon components such as carbon dioxide, nitrogen, hydrogen sulfide, helium, etc; however, in estimating potential supply, it is not feasible to separate small volumes of these components from the hydrocarbons. Areas or formations that are believed to contain large volumes of such nonhydrocarbon components are not counted as part of the potential supply.

Definition of Potential Gas

The phrase *potential supply of natural gas*, as used by the Potential Gas Committee in making its estimates and in preparing its report, means: At a given date and underlying a particular geographic area, that prospective quantity of natural gas yet to be found and proved (as the term *proved* is used by the American Gas Association Committee on Natural Gas Reserves) by all wells which may be drilled in the future under assumed conditions of adequate but reasonable prices and normal improvements in technology.

The definition of *potential supply* specifies a relationship to proved reserves because the Committee's estimate at any given date is not to include any proved reserves existing as of that date but is to include such volumes as may become proved reserves in the future. Potential supply is a volume of gas which is in addition to existing proved reserves. An estimate of potential supply must take into consideration the criteria used by the Committee on Natural Gas Reserves of the American Gas Association in preparing its annual estimates of proved recoverable reserves. The Committee on Natural Gas Reserves defines proved recoverable reserves (AGA, 1973, p. 102) as follows:

> The current estimated quantity of natural gas and natural gas liquids which analysis of geologic and engineering data demonstrates with reasonable certainty to be recoverable in the future from known oil and gas reservoirs under existing economic and operating conditions. Reservoirs are considered proved that have demonstrated the ability to produce by either actual production or conclusive formation test.
>
> The area of a reservoir considered proved is that portion delineated by drilling and defined by gas-oil, gas-water contacts or limited by structural deformation or lenticularity of the reservoir. In the absence of fluid contacts, the lowest known structural occurrence of hydrocarbons controls the proved limits of the reservoir. The proved area of a reservoir may also include the adjoining portions not delineated by drilling but which can be evaluated as economically productive on the basis of geological and engineering data available at the time the estimate is made. Therefore, the reserves reported by the Committee include total proved reserves which may be in either the drilled or the undrilled portions of the field or reservoir.

The Committee on Natural Gas Reserves includes in proved reserves all gas estimated to be producible from tested formations under existing operating and economic conditions without regard to the size, use, or disposition of any production. Proved reserves in an undrilled area, however, must be so related to the developed or tested leases and to known field geology that the areas productive ability is assured.

Categories of Potential Gas Supply

Accuracy of the estimates of gas volumes included in the potential supply of a given area are dependent upon geologic conditions and the extent to which the area has been explored and developed. Using available geologic data and The American Association of Petroleum Geologists' classification of wells (Fig. 1), the Work Committee divides the estimates into three broad categories. These categories are:

A. Probable potential gas supply (associated with existing fields).
   1. Supply from known accumulations obtained by:
      a. Future extensions of *existing pools,* in known productive reservoirs.
      b. Future *new-pool* discoveries, within existing fields, in reservoirs productive elsewhere within the same field.
   2. Supply from *new-pool* discoveries obtained by:
      a. Future shallower and/or deeper new-pool discoveries, within existing fields, in formations productive elsewhere within the same geologic province or subprovince, under *similar* geologic conditions.
      b. Future shallower and/or deeper new-pool discoveries, within existing fields, in formations productive elsewhere within

the same geologic province or subprovince, under *different*
geologic conditions.

B. Possible potential gas supply (associated with productive formations).
    1. Supply from *new-field* discoveries obtained by:
        a. Future new-field discoveries, in formations *productive* else-
            where within the same geologic province or subprovince, under
            similar geologic conditions.

C. Speculative potential gas supply (associated with nonproductive
    formations).
    1. Supply from *new-pool* discoveries in formations not previously
        productive within a productive geologic province or subprovince.
    2. Supply from *new-field* discoveries obtained by:
        a. Future new-field discoveries in formations *not previously*
            productive within a productive geologic province or subprov-
            ince.
        b. Future new-field discoveries within a geologic province *not*
            previously productive.

   A *geologic province* is defined as a "Large region characterized by
similar geologic history and development" (Weller, 1960, p. 26). Hence,
the Gulf Coast geosyncline and the Appalachian geosyncline commonly are
referred to as "provinces." Large provinces, such as those cited, are
divided into subprovinces to recognize geologic homogeneity. Examples
of subprovinces are the Mississippi embayment of the Gulf Coast geosyn-

DRILLING OBJECTIVE			ILLUSTR.	INITIAL WELL CLASSIFICATION	FINAL WELL CLASSIFICATION	
					SUCCESSFUL	UNSUCCESSFUL
EXPLOITATION OF ACCUMULATION DISCOVERED BY PREVIOUS DRILLING			1	DEVELOPMENT	DEVELOPMENT	DRY DEVELOPMENT
LONG EXTENSION OF PARTLY DEVELOPED POOL			2	OUTPOST OR EXTENSION	EXTENSION (SOMETIMES NEW-POOL DISCOVERY WELL)	DRY EXTENSION
NEW POOL ON STRUCTURE/ENVIRONMENT ALREADY PRODUCTIVE	WITHIN LIMITS OF PROVEN POOL	ABOVE DEEPEST PROVEN POOL	3	SHALLOWER POOL	SHALLOWER POOL DISCOVERY	DRY SHALLOWER POOL
		BELOW DEEPEST PROVEN POOL	4	DEEPER POOL	DEEPER POOL DISCOVERY	DRY DEEPER POOL
	OUTSIDE LIMITS OF PROVEN POOL		5	NEW-POOL WILDCAT	NEW-POOL DISCOVERY (SOMETIMES EXTENSION WELL)	DRY NEW POOL
NEW FIELD ON STRUCTURE/ENVIRONMENT NEVER BEFORE PRODUCTIVE			6	NEW-FIELD WILDCAT	NEW-FIELD DISCOVERY	DRY NEW-FIELD WILDCAT

known productive limits of proven pool

abridged from Dix and Van Dyke, AAPG Bull., vol. 53/6, p.1156, June 1969

FIG. 1—Classification of wells.

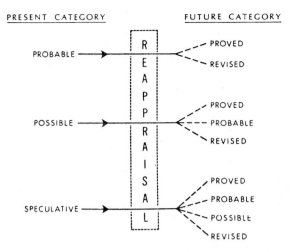

FIG. 2—Flow chart illustrating progression of potential gas supplies as result of drilling.

cline (province) and the Delaware basin within the Permian basin (province).

The term "basin" is avoided except in the preceding illustrations, because it has topographic and geomorphic meanings, as well as geologic. These meanings may be different and can lead to misinterpretations.

It is evident from the above that, as drilling progresses, gas volumes estimated to be in a particular reservoir will move from one category to another as shown schematically in Figure 2. Since any projection of potential supply lacks the accuracy of the proved reserve figures, particularly in the speculative category, it must be recognized that these estimates of potential supply will also be subject to upward and downward revisions when new geologic and engineering data are provided by exploratory drilling. When gas is finally classified in the *proved* category, it will no longer be included in the estimated potential supply.

TECHNIQUE FOR ESTIMATING POTENTIAL SUPPLY OF NATURAL GAS

Basic Approach

The basic technique for estimating potential gas supply is to compare the factors that control known occurrences of gas with factors present in prospective areas. Known occurrences are expressed as the volume of natural gas ultimately recoverable per unit volume of reservoir rock within an adequately explored part of a geologic province. Such known relationship, appropriately adjusted for variations in geologic and reservoir conditions, is then attributed to incompletely explored sedimentary rocks in the same or in a similar geologic province.

A general outline of the *attribution technique* for estimating potential supply of natural gas follows.

For Productive Formations

A. Within a productive province or subprovince, estimate the volume of productive gas-bearing rock and potential gas-bearing rock:

1. Associated with existing fields for estimating probable potential supply, and

    2. Associated with productive formations for estimating possible
       potential supply.

B. Add cumulative production and proved reserves to obtain the total
    volume of ultimate recoverable gas for the adequately explored por-
    tion of the productive gas-bearing rocks.

C. Divide this figure by the volume of adequately explored, productive
    gas-bearing rocks to obtain the ultimate recoverable gas per unit
    volume of productive rocks.

D. Adjust the unit volume figure for variations in geologic and reservoir
    conditions in the probable and possible gas-bearing rocks.

E. Using these adjusted figures:
    1. Estimate *probable potential supply* in extensions and new pools
       associated with existing fields, and
    2. Estimate *possible potential supply* by applying these adjusted
       figures to the wildcat traps and structures estimated to be pres-
       ent in the inadequately tested portion of the province or sub-
       province.

For Nonproductive Formations and Nonproductive Provinces

A. Estimate the volume of untested sediments in nonproductive provinces
    and the volume of potential but nonproductive sediments in productive
    provinces.
    1. Estimate the *speculative potential supply* in these sediments by
       comparing them with similar sediments in other provinces or sub-
       provinces where their productive characteristics are known.

Role of Judgment in Estimating Potential Supply

In proceeding from the known to the unknown, the judgment of the
estimator is the most significant factor in making estimates of potential
supply, particularly in the categories of *possible* and *speculative* supply. The appropriate adjustments are direct reflections of the estimator's
judgment. Only the estimator has the detailed knowledge and the experience necessary to select appropriate adjustments for the geologic provinces for which he is responsible. In all respects, the estimates of
potential supply by the Potential Gas Committee reflect an objective,
scientific approach to the problem. An attempt is made to use all pertinent geologic and engineering data.

Limiting Considerations in Making Estimates

At present, no potential supply of gas is considered at depths greater
than 30,000 ft (9,144 m) or, in offshore areas, where water depths are in
excess of 1,500 ft (457 m).

Economic, technologic, and governmental policy considerations also
impose restrictions in estimating potential supply, and to that extent
they are considered limitations of the estimates.

Economic and Technologic Aspects
of Estimating Potential Supply

Economic, technologic, and governmental policy considerations that
are taken into account in the Potential Gas Committee's estimate of potential gas supply are related to (1) past production and proved reserves of
natural gas, and (2) all wells which would be drilled in the future under
the assumed conditions that there will be adequate but reasonable prices
for gas and oil and normal improvements in technology.

Fundamental economic considerations—which include but are not limited
to prices, costs, rates of take, recovery factors, abandonment pressures,

etc—are inherent in the definition of *proved reserves*: "...recoverable in the future from known oil and gas reservoirs under existing economic and operating conditions." Some gas that is actually present in known reservoirs is not producible because of the relation between costs and prices, and therefore is not included in recoverable reserves. Furthermore, there are other known gas accumulations of such small size that none of the gas can be produced in commercial quantities under "...existing economic and operating conditions." Such considerations result in proved reserves being something less than the total volumes of gas existing in known reservoirs. These same limitations are applicable to the estimates of potential supply. If a fundamental change in economics or technology occurs, estimates of potential supply will be changed accordingly.

The attribution technique describes relationships for known occurrences of ultimately recoverable volumes of gas to prospective gas accumulations, after appropriate adjustment for variations in geologic and reservoir conditions are made. The explicit assumption of adequate but reasonable prices and normal improvements in technology in the definition of potential supply relates to improvements in exploration and production techniques. Adequate prices, normal technologic improvements, and reasonable governmental policies are required to bring about the drilling necessary to prove the potential supply. These assumed conditions permit estimates of potential supply to be made by the Potential Gas Committee on the basis of relevant past history and experience concerning recovery factors, as well as the size and type of reservoirs which have been found, developed, and produced, without speculating as to future levels of prices and costs.

REFERENCES CITED

American Gas Association, Committee on Natural Gas Reserves, 1973, Reserves of crude oil, natural gas liquids, and natural gas in the United States and Canada, and United States productive capacity as of December 31, 1972: Am. Gas Assoc., Am. Petroleum Inst., Canadian Petroleum Assoc., v. 27, p. 89-251.

Dix, F. A., and L. H. Van Dyke, 1969, North American drilling activity in 1968: AAPG Bull., v. 53, No. 6, p. 1151-1180.

Weller, J. M., 1960, Supplement to the glossary of geology and related sciences: Am. Geol. Inst., 72 p.

Undiscovered or Undeveloped Crude Oil "Resources" and National Energy Strategies[1]

EARL COOK[2]

ABSTRACT   Estimates of ultimately recoverable crude oil in the United States published within the past 10 years range rather widely. The highest estimate of recoverable oil remaining to be discovered is 15 times the lowest estimate. This spread is serious because national strategy based on the highest estimate could be quite different from that based on the lowest estimate.

If a high estimate is accepted, the nation might be justified in seeking to subsidize domestic discovery as an alternative to foreign imports and to development of substitute fuel systems. If a low estimate is accepted, appropriate strategies would be to subsidize the rapid development of substitute fuel systems, to build up a strategic economic reserve of petroleum, and to engage in government-to-government negotiations in order to try to assure a continued inflow of foreign oil at prices equal to or below that of substitute fuels.

Poorly defined terms and unjustifiable usages of figures representing a wide range of uncertainty are barriers to general understanding of fossil-energy futures. Geologic estimates of oil in place tend to project past costs of exploitation and to ignore exponential increases of work cost with depth and with reservoir recalcitrance; they also ignore the probability that "substitution" technology will outpace petroleum technology, and will transform most "undiscovered reserves," if they exist, into mere geologic anomalies. It may be that nongeologic methods of estimating future availability of oil and gas are better guides to national policy than are geologic methods.

ESTIMATE RANGE AND NATIONAL STRATEGY

Publications within the past 10 years on the amount of ultimately recoverable crude oil in the United States (Table 1) state or imply that the highest estimate of recoverable oil remaining to be discovered is 15 times greater than the lowest estimate. Such a range is almost useless for formulating national energy policy, because strategies based on the lowest estimate would be quite different from those based on the highest estimate.

Estimates of crude oil remaining to be added to reserves are calculated in two ways. One is by geologic analogy and geographic extrapolation of finding and recovery rates; the other is by statistical study and projection of the exploitation history. The former is the only practicable method for estimating the quantitative production potential of geologic terranes in which there has been little or no production; the latter method requires that the production history of the area of concern be beyond the youthful phase.

For purposes of national policy and strategies, the exploitation-history method offers a major advantage over the geologic-analogy method, because the former method expresses its conclusions, or can readily be made to do so, in years as well as in barrels. To the nation, knowing how much time there may be to prepare for the massive effort of replacing crude oil as a primary energy source and how many years the national depletion curve is ahead of the global depletion curve is more important than knowing how many barrels remain to be added to reserves.

CONCEPT OF NONECONOMIC "RESOURCES"

For many years a natural resource was regarded as a natural material or energy flow which could be exploited at a profit. As late as 1965,

[1]Manuscript received, January 6, 1975. Some of the material in this paper appears in *Technology Review*, June 1975.
[2]Texas A&M University, College Station, Texas 77843.

TABLE 1. ESTIMATES OF UNITED STATES CRUDE OIL RECOVERABLE BEYOND KNOWN RESERVES, 1965–1974 (BILLION BBL)

Author or Source	Year	Method	Oil in Place Initially Discoverable	Oil in Place Initially Discoverable	Ultimately Recoverable	Recoverable Beyond Known Reserves[1]
McKelvey and Duncan	1965	Geologic analogy	—	—	320/660[2]	184/524[2]
Hendricks	1965	Geologic analogy	1,600	1,000	400[3]	264
Weeks	1965	Geologic analogy	—	—	270	134
Hendricks and Schweinfurth	1966	Geologic analogy	2,000	1,250	500	364
Hubbert	1967	Exploitation history	—	—	170	34
Elliott and Linden	1968	Exploitation history	—	—	450	314
Hubbert	1969	Modified exploitation history	—	—	190[4]	54
Arps, Mortada, and Smith	1970	Exploitation history	—	—	165[5]	29[5]
Moore	1971	Exploitation history	—	587	353[6]	217
National Petroleum Council	1971	Geologic analogy	727	727	242[7]	106
Cram	1971	Geologic analogy	720	720	432[6]	296
U.S. Department of Interior	1972	Geologic analogy	2,830	—	549	413
Theobald, Schweinfurth, and Duncan	1972	Geologic analogy	—	1,895	596[8]	439[8]
Berg, Calhoun, and Whiting	1974	Modified geologic analogy	—	—	400	264
Ford Foundation Energy Policy Report	1974	Geologic analogy?	—	—	628[9]	466
Mobil Oil Corporation (in Gillette)	1974	Geologic analogy	—	—	—	88
Hubbert	1974	Modified exploitation history	—	—	213[10]	77

1 As of January 1, 1973. Cumulative production (100 billion bbl) and proved reserves (36 billion bbl) are not included.
2 Higher figure includes "resources" considered uneconomic in 1965.
3 Based on 40 percent cumulative recovery.
4 Includes 25 billion bbl for Alaska.
5 Does not include Alaska.
6 Based on 60 percent cumulative recovery.
7 Based on 33.3 percent cumulative recovery.
8 Includes NGL.
9 Sources and method are not given; figures are converted from heat-content equivalents of report.
10 Includes 43 billion bbl for Alaska.

Weeks (p. 1686) wrote: "...nothing can be considered a reserve or a resource unless it can be produced and used profitably...." Concepts such as potential resources and inferred reserves were carefully defined and used with the appropriate qualifying adjectives. Probable or indicated reserves were known to involve a higher order of uncertainty than proved or measured reserves; possible or inferred reserves were still another order removed; and speculative reserves were the most uncertain of all. Reserves in different categories of uncertainty were not added together—at least not by those who understood that adding together 1 partridge in a cage, 10 partridges in a pear tree, and 100 partridges that might be somewhere in the forest does not result in 111 partridges to eat.

As early as 1949, however, Sam Lasky, then of the U.S. Geological Survey, had suggested that the profitability constraint be reomved from the term "resources" and had defined resources as equal to the sum of (1) reserves, (2) marginal resources, (3) submarginal resources, and (4) latent resources, although he did not indicate how one would measure or determine the last three categories. Lasky (1949, p. 36) wrote:

> *Mineral resources* include all the material in the ground, discovered or undiscovered, usable at present or not, rich or lean, considered within the context of all factors, quantitative and qualitative, that may influence its conversion into a "reserve," and within the context of all factors that enter into prediction or opinion as to possible future usability.

From this subjective and essentially indeterminate concept of resources (see also Blondel and Lasky, 1956), it was but a step to the resource-base model.

RESOURCE-BASE CONCEPT

Economists of Resources for the Future, Inc., and geologists of the U.S. Geological Survey have promoted the use of the resource-base concept. The "resource base" includes "the sum total of a mineral raw material present in the earth's crust within a given geographical area" (Schurr and Netschert, 1960, p. 297), whereas "resources" are "the natural stock of the mineral raw material from which will come the supply of the metal [or mineral fuel] within the period considered" (Landsberg et al, 1963, p. 425). The audacity of such notions—in the face of the geologic complexities and our inability to see through rock, which have always plagued economic geologists, and in view of the difficulties of forecasting technology and prices in a market of increasing scarcity—is exceeded only by their potential for misleading those who will suffer when the ultimate recovery falls far short of the "resource base."

The resource base is a construct which demands that all "partridges," no matter how remote, elusive, or chimerical, be treated as equals. How else can one subtract from a quantitative vision of partridges initially in the woods ("initially" meaning when we found out they were good to eat) quantitative estimates of undiscoverable and inedible partridges, and then take away the quantitative amounts of past harvest and caged reserves, to arrive at an estimate of the partridges remaining to be eaten? I am speaking, of course, of "nonrenewable" partridges.

The pernicious impact of the resource-base concept, injudiciously presented, can be illustrated from two publications. The first report (U.S. Dept. Interior, 1968, p. vii), which is clearly aimed at the interested layperson, compares estimates of crude oil, natural gas liquids, and natural gas *originally in place* within the exploitable jurisdiction

TABLE 2.   RESOURCE BASE IN NON SEQUITURS[1]

Total original oil and gas in place in the United States and its continental shelf to a water depth of 600 ft [183 m] is estimated to be:

	Originally in Place	Withdrawn to 1/1/68
Crude oil (billion bbl)	2,000	84
Natural gas liquids (billion bbl)	150	n.a.
Natural gas (Tcf)	5,000	332

The remaining petroleum resources of the United States are obviously adequate to support consumption for many years into the future. The real question is whether they can be located and produced at costs which permit them to compete with other energy sources.

[1]This table appears under the heading "The Resource Base" in the U.S. Department of the Interior (1968) report.

of the United States with cumulative domestic production of oil and gas through 1967 (Table 2). The first sentence below the table hammers home the point: "The remaining petroleum resources of the United States are obviously adequate to support consumption for many years into the future."

The contradiction in the next sentence, as to whether these "resources" will ever become *reserves*, is not obvious to the nonprofessional reader, and he may never reach the passage which might have jolted him. It reads: "The fact that we have X billion barrels of oil and Y trillion cubic feet of gas in the ground, however, says nothing at all about how much of these same quantities will eventually be found and put to use."

The second publication (U.S. Dept. Interior, 1972) shows, in histogram form (Fig. 1), estimates of the United States crude oil resource base. In this figure, past oil production, undiscoverable and unrecoverable oil, and everything between are added together, both graphically and numerically.

Early in 1974 the first report of the Ford Foundation's Energy Policy Project was published. The staff and advisory board for the project were liberally larded with economists; Resources for the Future (RFF) was contracted to provide an analysis of future national energy supply. The table of United States energy resources in the project's first report is in resource-base format (Table 3); when converted to barrels from the heat equivalents given, the figures for petroleum represent new highs in estimates of oil originally in place ($3,680 \times 10^9$ bbl) and in "recoverable resources" (520×10^9 bbl). Not surprisingly, the report states (p. 44): "The work done for the Project by RFF suggests that energy resources are at least sufficient to meet the year 2000 requirements with major reliance on oil and gas supply."

The use of the term "resource" in the ways documented above can mislead the uninformed into unjustified complacency about the future availability of nonrenewable geologic commodities. It encourages a picture of an inevitable continuity from potential resource to proved reserve to usable commodity.

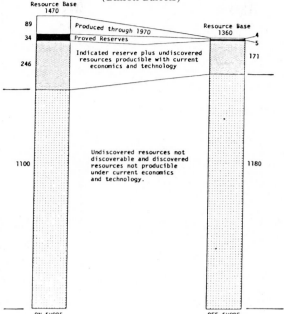

FIG. 1—A visual aid to error. From U.S. Department of the Interior (1972, p. 27).

However useful the resource-base concept is as a conceptual tool, its use outside the community which knows its inherent defects should be attended with great care, because it is much too comforting. Any approach to a societal problem which carries within it such a potential for pandering to the common thirst for good news can be hazardous to society's future health. The resource-base concept and the high estimates of future availability it fosters are comforting to those already comfortable and reassuring to those who wish to become comfortable. They accord with the group interests and prejudices of economic geologists and of resource-exploiting organizations; they minister to faith in man's unlimited ingenuity (sometimes called "technology") and to the concomitant notion that man's economic laws transcend nature's physical laws; they lull the fear of scarcity and lighten the shadow of sacrifice. Estimates based on the resource-base concept should be viewed with extreme skepticism simply *because* they are so comforting.

However, there are other reasons for viewing the resource-base approach with disfavor. Geologic estimates of recoverable oil in place not only may carry analogy to extremes unjustified by experience, but they tend to project past work costs of exploitation and to depreciate or ignore exponential increases of work cost with depth (Fig. 2) and with reservoir recalcitrance. Estimates of oil in place likewise ignore the probability that "substitution" technology will outpace petroleum technology and transform most "undiscovered reserves," if they exist, into mere geologic anomalies. The inherent assumption of irreversibility of the "resource" flow from submarginal or paramarginal through marginal to reserves status must be based on faith, for it is not a conclusion

TABLE 3. A NEW HIGH IN "RECOVERABLE RESOURCES" OF PETROLEUM[1]
Major Energy Resources, United States

	1973 Consumption (Quadrillion Btu)	Cumulative Production (Q Btu)	Reserves[2] (Q Btu)	Recoverable Resources[2] (Q Btu)	Remaining Resource Base[2] (Q Btu)
Petroleum	34.7	605	302	2,910	16,790
Shale oil	—	—	(465)	N/A	975,000
Tar sands	—	—	—	N/A	168
Natural gas	23.6	405	300	2,470	6,800
Coal	13.5	810			
Strippable coal	N/A	N/A	4,110	14,600	64,000
Low-sulfur coal	N/A	N/A	925	2,600	2,600
Uranium			2,390	N/A	38,200
Used in light-water reactors	0.85	2	228	600	3,200
Used in breeders	—	—	17,700	47,000	200,000,000
Thorium used in breeders	—	—	4,200	17,500	570,000
Hydropower	2.9	—	—	5.8[3]	—

$$2,910 = 520 \times 10^9 \text{ bbl at } 5.6 \times 10^6 \text{ Btu/bbl}$$
$$-302$$
$$2,608 = 466 \times 10^9 \text{ bbl at } 5.6 \times 10^6 \text{ Btu/bbl} = \text{Recoverable resources remaining to be discovered or added to reserves}$$

[1] This table is modified from Ford Foundation Energy Policy Project (1974).
[2] These are geologic estimates. Reserve estimates are based on detailed geologic evidence, usually obtained through drilling; the other estimates reflect less detailed knowledge and more geologic inference. All estimates are based on assumptions about technology and economics. They may increase over time as technology improves or prices increase.
[3] Ultimate capability.
N/A = not available.

that can be sustained by logic; the moment it costs more to find and produce a barrel of "new" crude oil than it does to manufacture a barrel of substitute crude oil from coal, oil shale, or "tar" sands, there will be no more undiscovered crude oil "resources" and the question of how much oil remains in the ground will be of no further importance.

It does no good to rely on the unlimited ingenuity of man to overcome physical laws. Man's ingenuity may indeed be unlimited, but his ability to apply the products of his ingenuity economically *is* limited.

The statistical basis for rejecting the resource-base model as a guide to estimation of future recovery has been thoroughly presented by Hubbert (1974), who pointed out the "leverage" on estimates of remaining recoverable supplies represented by changes in estimates of the base or of the ultimate recovery ratio. The recovery ratio, as a manipulatable factor in projection based on oil-in-place estimates, unbridles the technological enthusiast and allows estimates that put extreme demands on the technological cavalry or on Providence.

The resource-base model fits the fiction of the "endlessly retreating line" (Merrill, 1959, p. 36) that is supposed to divide the profitably minable from submarginal materials. This cornucopian thesis has been epitomized by economist Carl Kaysen (1972, p. 661, 663), who wrote: "The fact that some limits exist, that the earth is in principle finite, is hard to deny, but does not in itself lead to any very interesting conclusions....Resources are properly measured in economic, not physical terms."

Some deposits of some mineral resources seem to fit the "endlessly retreating line" hypothesis; most do not. Among those minerals whose deposits do not seem to fit at all are silver, mercury, and crude oil. Deposits of each are typically small and have sharp boundaries. In

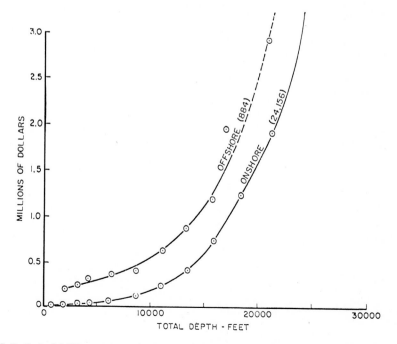

FIG. 2—Cost of drilling and equipping oil and gas wells (including dry holes) in United States, 1971. Based on American Petroleum Institute et al (1972).

addition, the larger deposits of each appear to be concentrated near the earth's surface. The economic significance of these three observed characteristics is that production in the later stages of depletion of a district or a continent is not a linear function of price. The price and production histories of silver and mercury in the United States demonstrate clearly that, once the production peak has been passed, rises in price excite weaker and weaker responses in production. This same relation is about to be demonstrated for crude oil and natural gas.

EFFECT OF HIGH AND LOW RESOURCE ESTIMATES

We already have experienced one of the expectable political conse- quences of large estimates of "recoverable resources"—a lack of belief in published proved-reserve figures and a pervasive suspicion of con- trivance in shortages of petroleum products. When it becomes evident that very large price rises for oil and gas, as well as continuing "tax breaks," are not producing anything like an equivalent return in new production, this suspicion may become an uncontrollable political force that will sweep the domestic oil and gas industry into the dustbin of free enterprise.

That large "resource" estimates may dwindle to small additions to reserves is only one reason large estimates are a poor guide to national policy. Even if such an estimate should be verified by production, the amount of additional time the resource will be available may be increased relatively little. A resource that is available will be used. If avail- ability increases, consumption will increase; when consumption increases, depletion is increased. Hubbert's (1969) logistic curves (Fig. 3) illus- trate this point, as do calculations by Elliott and Turner (1972), which were made by several different methods for all the world's fossil fuels.

A high estimate of the domestic quantity available (which may be translated as "a long time" by national decisionmakers!) may lead to a set of national strategies featuring incentives for domestic exploration and production, relaxation of environmental constraints on such activities,

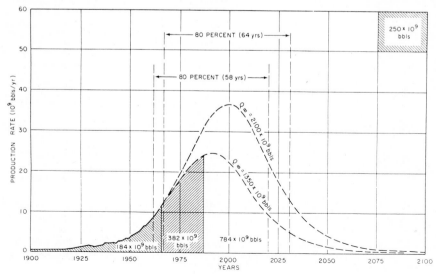

FIG. 3—Complete cycle of world crude oil production. From Hubbert (1969) by permission of the National Academy of Sciences.

and restrictions on imports in order to encourage domestic exploration. On the other hand, a low estimate of the quantity available from domestic sources—or of the time remaining for emplacement of an adequate substitute system—may lead to different national strategies highlighted by subsidized development of coal conversion, nuclear technology, and shale-oil production; strong conservation incentives; creation of a strategic economic reserve or stockpile; continued efforts to secure foreign resources, and, possibly, disincentives to domestic exploration and production in order to save crude oil for a higher use or a greater need. The two sets of strategies, and the policies they would be designed to implement, would differ substantially, as would their potential benefits and costs to the nation.

Because time available is more critical to national welfare than quantity available, it may be that a nongeologic method of estimating future availability of oil and gas based on exploitation history is a better guide to national policy than are geologic methods. The exploitation-history method tends to restrain subjective influence and is expressed in rates as well as quantities. It should be noted, however, that geologic methods can result in low estimates and that exploitation-history studies have resulted in high estimates. The critical factor, then, may not be in the method but in the logic used.

One may thrill to the rally call of a Thane McCulloh (1973)—who proclaimed that, "To discover petroleum and new petroleum provinces requires an open mind, a desire to explore, a positive and adventurous attitude, a gambling spirit, and a will to win...."—without being willing to commit the future of the country to such gamblers.

REFERENCES CITED

Arps, J. J., M. Mortada, and A. E. Smith, 1971, Relationship between proved reserves and exploratory effort: Jour. Petroleum Technology, June 1971, p. 671-675.

Berg, R. R., J. C. Calhoun, Jr., and R. L. Whiting, 1974, Prognosis for expanded U.S. production of crude oil: Science, v. 184, p. 331-336.

Blondel, F., and S. G. Lasky, 1956, Mineral reserves and mineral resources: Econ. Geology, v. 51, p. 686-697.

Cram, I. H., 1971, Future petroleum provinces of the United States—their geology and potential—summary, p. 1-54: AAPG Mem. 15, 2 vols., 1495 p.

Elliott, M. A., and H. R. Linden, 1968, A new analysis of U.S. natural gas supplies: Jour. Petroleum Technology, February 1968, p. 135-141.

——and N. C. Turner, 1972, Estimating the future rate of production of the world's fossil fuels: Paper presented at Am. Chemical Soc. Mtg., Boston, April 9-14, 21 p.

Ford Foundation Energy Policy Project, 1974, Exploring energy choices: Washington, D.C., 81 p.

Gillette, Robert, 1974, Oil and gas resources: Did USGS gush too high?: Science, v. 185, p. 127-130.

Hendricks, T. A., 1965, Resources of oil, gas, and natural-gas liquids in the United States and the world: U.S. Geol. Survey Circ. 522, 20 p.

——and S. P. Schweinfurth, 1966, Unpublished memorandum, cited *in* United States petroleum through 1980: U.S. Department Interior, 1968, p. 11.

Hubbert, M. K., 1962, Energy resources: Natl. Acad. Sci. - Natl. Research Council Pub. 1000-D, 141 p.

——1967, Degree of advancement of petroleum exploration in the United States: AAPG Bull., v. 51, p. 2207-2227.

——1969, Energy resources, *in* Resources and man: San Francisco, W. H.
   Freeman, p. 157-242.

——1972, Survey of world energy resources: Paper presented to 26th Ann.
   Conf. Middle East Institute, Washington, D.C., September 29-30, 37 p.

——1974, U.S. energy resources, a review as of 1972: U.S. Senate Comm.
   Interior and Insular Affairs, 93d Cong., 2d Sess., Comm. Print Serial
   No. 93-40, 267 p.

Kaysen, Carl, 1972, The computer that printed out W*O*L*F*: Foreign
   Affairs, v. 50, p. 660-668.

Landsberg, H. H., L. L. Fischman, and J. L. Fisher, 1963, Resources in
   America's future: Baltimore, Johns Hopkins Press, 1017 p.

Lasky, S. G., 1949, National mineral resource appraisal: Mining Congress
   Jour., January, p. 35-37.

McCulloh, T. H., 1973, Oil and gas, *in* United States mineral resources:
   U.S. Geol. Survey Prof. Paper 820, p. 477-496.

McKelvey, V. E., and D. C. Duncan, 1965, United States and world resources
   of energy, *in* Symposium on fuel and energy economics: 149th Natl.
   Mtg., Am. Chemical Soc., Div. Fuel Chemistry, v. 9, no. 2, p. 1-17.

Merrill, C. W., 1959, The significance of the mineral industries in the
   economy, *in* Economics of the mineral industries: New York, AIME, p.
   1-42.

Moore, C. L., 1971, Analysis and projection of historic patterns of U.S.
   crude oil and natural gas, *in* Future petroleum provinces of the
   United States—their geology and potential: AAPG Mem. 15, p. 50-54.

National Petroleum Council, 1971, U.S. energy outlook—an initial
   appraisal, 1971-1985: Washington, D.C., v. 2, 195 p.

Schurr, S. H., and B. C. Netschert, 1960, Energy in the American economy,
   1850-1975: Baltimore, Johns Hopkins Press, 774 p.

Theobald, P. K., S. P. Schweinfurth, and D. C. Duncan, 1972, Energy
   resources of the United States: U.S. Geol. Survey Circ. 650, 27 p.

U.S. Department of the Interior, 1968, United States petroleum through
   1980: Washington, D.C., Office of Oil and Gas, 92 p.

____1972, U.S. energy—a summary review: Washington, D.C., U.S. Govt.
   Printing Office, 42 p.

Weeks, L. G., 1965, World offshore petroleum resources: AAPG Bull.,
   v. 49, p. 1680-1693.

Assisting Project Independence—A National Program [1]

JOHN L. STOUT [2]

ABSTRACT   Studies and reevaluations of petroleum resources made recently indicate that full self-sufficiency for domestic natural resources by 1980 is unlikely. Most investigators believe some measure of independence must be achieved by 1985. By 1985 our oil imports may increase to one half of our consumption, in comparison with one third at the present time. A recent article pointed out that United States self-sufficiency would require a $310 billion investment during the 12 years 1974-1984; however, costs of deeper drilling and continued inflation probably will increase this figure to $500 billion.

Petroleum Information's data and experience have been matched with part of the economic model developed by Lawrence Livermore Laboratory and published as the Conference Board's prediction of U.S. future energy requirements. During the past 20 years, the annual drilling rate reached a maximum of over 57,000 wells in 1956 and a minimum of 26,328 wells in 1971. Oil-well completions were highest in 1956 (31,000 wells) and lowest in 1973 (9,000 wells). Only one eighth of the wells completed after gas price regulations became effective were completed as gas wells, but one fourth of the wells completed in 1974, with new gas price incentives, were gas wells. In order to maintain our present ratio of reserves to production, we should be completing twice as many wells as the expected total for 1974. Furthermore, we must sustain that higher rate of completion each year through 1985.

Model studies necessary for economic evaluation of energy resources require information data banks such as the one available at Petroleum Information. Our proposed study would be made by geologic provinces within geographic areas. It would include reserves information, economic data, statistical evaluations, and other information; 6-month updates would be made. Evaluations would be made by use of a probability model which considers exploration as a sampling process. Benefits of the project would be revealed in the consistent figures for proved oil and gas reserves, sufficiently detailed for economic study.

INTRODUCTION

Since 1973, when President Nixon announced "Project Independence," there have been repeated studies and reevaluations of United States petroleum resources. All arrive at the same conclusion—that full self-sufficiency for domestic natural resources by 1980 is unlikely. Most investigators agree, however, that some measure of independence must be achieved by 1985. Economic repercussions will be severe if we do not become more independent of our present sources of energy.

It is obvious from all forecasts that oil and gas will continue as our major source of energy until we can develop practical production of other sources such as nuclear and mineral fuels or geothermal and solar energy. In the meantime, oil will supply over one half of this country's rapidly increasing energy needs, and by 1985 our oil imports may increase to one half of our consumption. At present, we import one third of the oil consumed. "Even at the old prices a few months ago, the cost of such an import level would be $25 billion annually, an impossible burden on the U.S. balance of payments. . . . At the new prices, the costs would be astronomical—perhaps prohibitive," according to David Rockefeller (1974), who spoke to the Centennial Convocation of the Colorado School of Mines.

[1] Manuscript received, January 31, 1975.
[2] Petroleum Information Corporation, Denver, Colorado 80201.

INDUSTRY INCENTIVE

A change of incentive is necessary to reverse the declining exploration trend this country has experienced. The *Journal of Petroleum Technology* (1974) points out that, during a 10-year period (1964-1973), the
domestic petroleum investment totaled $86 billion, whereas the U.S.
petroleum industry profit totaled $26 billion—a deficit of $60 billion!
The article concludes that, "During the 12 years between 1974 and 1984
[inclusive]. . . U.S. self-sufficiency will require a $310 billion invest-
ment."

INDUSTRY PROFILE STUDIES

Petroleum Information Corporation's major business experience over
the last 46 years has been the accumulation of accurate and comprehensive
data on the petroleum industry. The 31 offices throughout the United
States are staffed with trained professionals accustomed to interpreting
the significance of industry data. We have applied our experience to the
economic model developed by Lawrence Livermore Laboratory and published
as the Conference Board's prediction of U.S. future energy requirements
(Murphy, 1972). The Conference Board is an independent, nonprofit,
business-research organization. For over 50 years, their 4,000 worldwide
associates have been dedicated to scientific research in business economics. This paper is Petroleum Information's evaluation of past industry
conditions and future requirements.

PAST INDUSTRY CONDITIONS

During the past 20 years, 1956 was the year of maximum drilling with
just over 57,000 wells completed. The annual drilling rate declined
steadily to a minimum of 26,328 wells completed in 1971. The number of
active drilling rigs was also at its lowest in 1971 (1,025 rigs) and,
according to the *Journal of Petroleum Technology* (1974), the industry
rate of return on net worth was 11 percent. This was just one percentage
point above the rate of return for the heavily government-regulated
commercial banks.
There has been a significant change in the ratio of oil and gas wells
completed. Oil-well completions were highest in 1956 (31,000 wells put
on production) and lowest in 1973, when only 9,000 wells were completed
as oil producers. Only one eighth of the wells completed after gas price
regulations became effective were completed as gas wells, but, with new
gas price incentives, one fourth of the wells completed in 1974 were gas
wells. There were 6,300 gas wells completed in 1973 compared with 3,800
in 1967, one of the lowest years.

FUTURE INDUSTRY REQUIREMENTS

In order to meet future demands for energy, the Conference Board's
report (Murphy, 1972) on energy and public policy predicted that we must
sustain just over 4 billion bbl of oil production annually and increase
gas production from 22 Tcf to 24 Tcf per year by 1976. Gas production
could decline to 20 Tcf annually by 1985.
If we hope to maintain our present ratio of reserves to production, we
should be completing twice as many wells as the expected 30,000 wells in
1974; such a drilling rate would equal the industry's highest performance

in the past 20 years. Also, we must sustain that higher rate of completion each year through 1985. At that rate, we will have completed 685,000

NATIONAL STUDY OF RESERVES

The essential research and development suggested in this paper are
designed to meet the nation's need for expanded supplies of energy.
Investigators of energy reserves include government agencies, congressional
committees, petroleum producers and marketers, and many research institutions. We believe that the already-compiled information systems on wells
and production (Petroleum Information Corporation has accumulated a data
bank of over 755,000 well histories and a production data bank of three
quarters of all the oil and gas production in the United States) should
be utilized in fundamental research bearing both on short-term solutions
to pressing problems and on long-range planning. These data banks can be
interrogated by uniquely qualified, independent technicians to prepare
industry statistics.

Use of this information data system would avoid duplication of effort
and waste of research funds to reconstruct information already available.
It is believed that there are, for example, sufficient prospect details
in the present data bank to assist prudent investment of $125 million to
drill 5.5 million ft (1.7 million m) to explore for gas, hopefully adding
1 Tcf of gas to our annual domestic production. Our data would be particularly significant for the first 2 years of "Project Independence."
Regular updates at 6-month intervals are suggested to be of value in
monitoring domestic-reserves progress.
wells in the next 12 years (303,360 oil wells, 61,730 gas wells, and
the balance will be dry holes). This completion rate is based on an
average deliverability of 172,000 bbl of oil and 4.1 Bcf of gas for each
productive well. The cost of deeper wells and continued inflation will
probably add $10 billion per year, or up to $500 billion investment, just
to meet the predicted demands. The initial argument of Project Independence was to maintain the present ratio of reserves to production while
meeting the future energy requirements. This would indicate that the estimated $310 billion for self-sufficiency is unrealistically low.

Model Studies

Several very extensive and detailed models have been developed for
economic evaluation of energy resources. All require information data
banks for source material. An example is the development history of the
Denver-Julesburg basin of Nebraska and Colorado, conducted by Petroleum
Information. The details of this study, complete through September 1972,
related that 21,371 wells were completed during the basin development.
Of these, 12,238 were exploratory wells and 9,133 were development wells.
Of the exploratory wells, 1,265 resulted in oil discoveries and 192 were
gas discoveries, but not all were commercially productive. There have
been 820 fields developed, and the combined ultimate recovery for the
basin is 850 million bbl of oil.

Over one half of all wells drilled in the basin were completed before
1959, resulting in 338 oil fields with an average size of 100,000 bbl.
There was a marked drop in drilling activity after this initial development; a low of only 225 exploratory wells was reached in 1966. However,
drilling has increased to an average of 588 exploratory wells drilled
per year since 1968, and the average field size now is 2 million bbl.
Operators are now finding oil reserves valued at $5 for every $1 invested,
whereas in 1958 the return was barely equal to the investment.

In another study of gas evaluation in the northern Rockies, a basin-analysis technique was developed by Petroleum Information from isopach maps of five major stratigraphic zones. The study evaluated proved gas production in each interval. This study was derived by use of data from 10,000 exploratory and development wells with specific gas information, after searching the files on 50,000 wells recorded in the project area. Detailed maps posted with wildcat oil and gas shows within each stratigraphic zone were included in the final report. This study resulted in evaluation of the exploration success expressed in discovered reserves, and was used to derive statistics relative to the amount of hydrocarbons remaining in the geologic province. The ratio of explored and unexplored volume of reservoir rock was used to extrapolate the hydrocarbon potential.

Technical Aspects

Essential in any reservoir study is availability of digitized electronic well logs suitable for formation evaluation. Map and electronic-log display capabilities are also vitally important. Recent advances in these techniques include accurate representation of fault traces and determination of faulted-reservoir volumes, both of which are superior to the old planimeter method of evaluation.

An additional consideration of reservoir evaluation is availability of adequate data, based on monthly production of crude oil and gas. The credibility of facts, their completeness, and their accuracy are crucial to the integrity and overall value of the study. The fact that such data have been carefully compiled and checked objectively tends to guarantee ready acceptance of the studies by engineers. Reserves can be reliably calculated from the performance profile of the decline curve and from estimates of the reservoir parameters gained from log analysis. Also, if sufficient information on pressures and periodic production is available, a method of determining the amount of hydrocarbons in place can be calculated by use of the material-balance technique. The purpose of all these routines is to determine hydrocarbon volumes initially present in the reservoir. Results from these computations can be upgraded as new periodic data become available.

POTENTIAL APPLICATIONS

In December 1973, the American Gas Association (1974) submitted a research and development plan for 1974-2000. Their plan for the period 1974-1985 includes additional methods for estimating potential gas supplies as well as advanced theoretical means of understanding hydrocarbon formation, migration, and entrapment and geophysical methods of detecting gas potential. In their plan for exploration and production, they intend to:

1. Increase the natural gas discovery rate and improve recovery;
2. Reduce cost required for discovery and production of natural gas; and
3. Present improved utilization of existing information and future data generated on exploration and production technology.

Other topics under consideration for exploration and production research by the American Gas Association include the following:

1. Optimization of geologic exploration strategy and natural gas production;

2. Determination of the effect of gas prices on presently known gas
    reserves;
3. Prediction of domestic gas producing rates and estimation of gas
    reservoirs;
4. Prediction of gas demand using Monte Carlo simulation.

PROPOSED PROGRAM

In the projected study, the compilation would best be made by geologic
provinces within geographic areas. Where sufficient producing data are
available, the development wells and pool reserves will be tallied. Some
areas of past dense drilling where records are incomplete will require
estimates of development wells drilled and gross estimates of reserves.
In most newer areas, exploratory information is detailed enough to permit
processing of maps and gas shows indicating remaining gas reserves. The
project might include historic figures on expenditures for exploration,
development, and production. These details would be suitable for additional analysis by clerical or computer economic models.

Study Results

Suggested reports would include (1) producing-field names, year discovered, cumulative reserves, and estimated remaining reserves; (2) annual
expenditures, by year, for exploration, development, and production for
the last 10 years, where available; (3) statistical evaluation of basin
history by year and depth, including 5-year estimates of the future;
(4) regional maps of oil and gas fields posted with detailed information
on hydrocarbon shows; and (5) 6-month update with revisions and corrections to previous estimates.

Any evaluation of the data in the various phases of the project would
be done by use of a probability model which considers exploration as a
sampling process; each wildcat is considered an observation from a geologic basin containing an unknown number of reservoirs. Exploration in a
basin is sampling without replacement, since nature does not replace
producing fields upon discovery. The likelihood of finding large reservoirs diminishes with each discovery. This is a characteristic of exploration which explains why the success ratio in a particular geologic
province is high during the early stages of exploration and decreases
rapidly throughout the drilling history of the basin. Each basin must be
defined by its geologic environment to formulate the geologic hypothesis
regarding the hydrocarbon potential in each sedimentary basin.

Evaluation of the hydrocarbon and economic potential after past production must take into consideration the economic potential and current
pricing. This technique could be applied to all geologic provinces in
the United States or, where information is available, throughout the
world. The history of the exploratory drilling and estimates of proved
reserves are available or can be derived. It is also possible to calculate approximate drilling cost based on studies such as the annual publication by the American Petroleum Institute, "Joint Associations Survey of
the United States Oil and Gas Producing Industry."

Benefits

The benefits of this project, of course, would be revealed in the
consistent figures for proved oil and gas reserves, sufficiently detailed
for economic study. Petroleum Information believes this project could
be conducted as an objective review from standardized data. The accuracy

of the evaluation of the petroleum industry could be most valuable in
future positive planning for the industry. Once the basic information
of this project has been accumulated, the maintenance phases would then
be most useful in predicting the effectiveness of future drilling and
exploration expenditures. The proposed maintenance of information for
this project might well become the most beneficial part of the study,
but such maintenance depends on the sound establishment of the information phase derived from the previous phases of development. This study
could be the basis for other research and development studies of the
nation's future energy requirement.

REFERENCES CITED

American Gas Association, 1974, Gas industry research plan, 1974-2000:
      Executive Committee Brochure on Research and Development, No. M-20274,
      100 p.
Journal of Petroleum Technology, 1974, TIC facts: profits are essential
      for a solution to the energy "crisis": v. 26, June, p. 622.
Murphy, J. J., ed., 1973, Energy and public policy--1972: New York,
      Conference Board, Inc., A conference Report, 1972, 318 p.
Rockefeller, David, 1974, Living in an energy-scarce world, *in* Proceedings
      of the 1974 Centennial Conference and Convocation: Colorado School
      of Mines, p. 76-80.

A Probabilistic Model of Oil and Gas Discovery [1]

G. M. KAUFMAN,[2] Y. BALCER,[3] and D. KRUYT[4]

ABSTRACT   A probabilistic model was constructed of the size of pool discovered in order of discovery within a geologic zone. The model predicts a decline in the average size of discovery as the resource base is depleted. It is built on assumptions about the size distribution of hydrocarbon deposits and the way which this size distribution interacts with exploratory activities. These assumptions govern the behavior of additions to discovered oil (gas) in place as a function of wells drilled in a play. Statistical properties of major Alberta plays were compared with properties of a Monte Carlo simulation of the model. It is possible to interface the model with expert subjective judgment to generate probabilistic forecasts of the size distribution of pools in prospective areas.

INTRODUCTION

A coherent national energy policy cannot be formulated without reliable estimates of the quantities of oil and natural gas remaining to be discovered in United States territories, supplemented by a forecast of what fraction of each can be recovered using currently available technology. Unfortunately, there is wide disagreement about what methods should be used to generate these estimates, as well as about their magnitude: the highest publicly cited estimate of recoverable oil remaining to be discovered is about 17 times the lowest!

A national energy policy based on the lowest of these estimates may differ radically in form from one based on the highest. Therefore, it is of critical importance to develop methods for estimation of oil and gas reserves that have scientific credibility and that simultaneously generate estimates in a form immediately useful for policy analysis. Unfortunately, none of the methods currently employed to estimate amounts of undiscovered oil and gas recoverable by use of current technology possess both of these attributes. The primary purpose of the research program proposed here is to develop methods that possess both. In order to be scientifically credible, a method must be based on explicitly stated postulates whose validity can be empirically confirmed using observed data. In order to be useful for policy analysis, it must provide not only single-number estimates, but an explicit measure of the degree of uncertainty of each such estimate.

In addition, methods should be designed so as to allow construction of an economic supply function (i.e., a description of how additions to reserves from new discoveries behaves as a function of wellhead price, exploratory effort, and the costs of exploration). Our goal is the construction of a predictive model which provides probabilistic answers to two questions:

1. How many undiscovered pools remain in a given region, and what is their size distribution?
2. What additions to economically exploitable reserves will accrue from an increment of exploratory effort?

[1]Manuscript received, January 24, 1975. This work was supported by National Science Foundation Grant GI 34936.
[2]Alfred P. Sloan School of Management, Massachusetts Institute of Technology, Cambridge, Massachusetts 02139.
[3]Department of Economics, Massachusetts Institute of Technology, Cambridge, Massachusetts 02139.
[4]Department of Economics, Harvard University, New Haven, Connecticut 06520.

The model can be interfaced with expert subjective judgment to provide an answer to the first question for unexplored areas, as well as
for areas where data on drilling successes and failures and sizes of
discoveries have been generated by exploration activity.

It is a process-oriented probabilistic model. By "process-oriented"
we mean a model that explicitly incorporates certain geologic facts and,
in addition, is based on assumptions that describe the manner in which
exploration technology and observed statistical regularities of the size
of pools interact to generate discoveries.

The proposed model of the discovery process has four major components:

1. A submodel of pool sizes discovered in a homogeneous geologic
population of pools in order of discovery.
2. A submodel of wildcat drilling successes and failures.
3. A submodel of the economics of a single exploratory venture.
4. A submodel of the "capital market" for exploratory ventures.

When assembled, these submodels constitute a probabilistic model of
the returns in barrels of oil and/or Mcf of gas generated as a function
of price and physical nature of the reservoirs available for exploitation.
We shall discuss properties of only the first of these four components.

DISCOVERY PROCESS

"Discovery process" is a descriptive label for the sequence of information-gathering activities (e.g., surface reconnaissance and magnetic,
gravimetric, and seismic surveys) and acts (drilling of exploratory wells)
that culminate in the discovery of petroleum deposits. In building models
of the discovery process, we will regard it as being effectively described
by a small number of quantitative attributes (such as the number of exploratory wells drilled into a geologic formation in a given area and
the oil [gas] in place in a newly discovered pool) and postulated relations among them. Although doing descriptive injustice to the way in
which geologists extrapolate geologic facts to guide exploratory activity,
a model composed solely of such attributes can embody many of the essential features of the discovery process.

A petroleum basin or area the size of Alberta will in general contain
reservoirs or pools with distinctly different geologic characteristics.
We shall regard the totality of pools in Alberta as being classified
into a collection of subpopulations of pools of similar geologic type.
By definition, a *play* begins with the exploratory well that discovers
the first pool in a subpopulation of pools. Thus there are, in principle,
as many potential plays as subpopulations or geologic types. The choice
of typology depends on the use to which it will be put; our choice will
be coincident with a generally agreed-upon description of major plays in
Alberta (e.g., Cardium, D-2, D-3, Viking, Beaverhill Lake).

A key component of our model is a set of (probabilistic) assumptions
which govern the behavior of additions to oil (gas) in place as a function of the number of wells drilled in a play. When plays are set in
relation to one another on a time scale, *total* additions to oil (gas) in
place in a given time interval may be regarded as generated by a temporal
superposition of individual plays. One might also superpose plays on a
scale composed of the cumulative number of exploratory wells drilled in
the province. A model that effectively describes the behavior of the
number of exploratory wells drilled into each play in any given time
interval automatically generates a description on this scale.

To the degree that we can separate physical and engineering aspects
of the discovery process from economic considerations, we shall do so.

A partitioning of assumptions into two classes, one physical and the other economic in character, leads to substantial simplifications both in the structure of the model and in procedures for making inferences about its parameters. In particular, classification of pools into geologically homogeneous subpopulations leads to a corresponding statistical homogeneity of the economic attributes of pools within each subpopulation. Thus we are able to trace the influence of price, exploration costs, and development costs on additions to reserves from new discoveries in a much more meaningful way than if all subpopulations of pools are aggregated into a single population.

Assumptions about the physical nature of the discovery process are stated in a way which tacitly implies that economic variables may influence the temporal rate of drilling exploratory wells in a play, but they do not affect either the probability that a particular well will discover a pool or the size of a discovery *within a given play*. This assertion is patently false if applied to a population consisting of a mixture of subpopulations with widely varying geologic characteristics. For example, a large price rise may accelerate exploratory drilling in high-risk (low probability of success) subpopulations with large average-pool sizes at a substantially different rate than in subpopulations with small pool sizes but high success probabilities. The overall probability of success for a generic well among the wells drilled in a mixture of these subpopulation types, as well as the size of discovery, will depend on the relative proportions of wells drilled in each subpopulation; and these proportions are influenced by prices and costs. By contrast, it is reasonable to assume that, within a given subpopulation, the precision of information-gathering devices and the quality of geologic knowledge of that subpopulation are the principal (perhaps sole) determinants of the probability of success of a generic well. A price rise may accelerate the temporal rate of drilling within that subpopulation, but it will not affect the quality of geologic knowledge at any given point on a scale of cumulative wells drilled into it. Exceptions can be found, of course, but this assumption is plausible as a broad descriptive principle. As stated, its adoption yields important analytical bonuses: it simplifies the modeling process and allows us to be parsimonious in choice of parametric functions for components of the model.

PHYSICAL POSTULATES

Our postulates or assumptions about the physics of the unfolding of a play reflect both petroleum folklore and the content of a variety of statistical and analytical studies of the discovery process. The principal ones are:

I[a] The size distribution (in barrels or Mcf) of petroleum deposits in pools within a subpopulation is lognormal.

II Within a subpopulation, the probability that the "next" discovery will be of a given size (in barrels or Mcf) is equal, to the ratio of that size to the sum of sizes of as-yet-undiscovered pools within the subpopulation.

The assertion that the size distribution of pools is lognormal implies that an individual accumulation, no matter how small, may be regarded as a pool. The number of such pools in a given play can be enormous. Since "tiny" accumulations are of no practical importance, we can, a priori, restrict the definition of elements of a subpopulation of pools to include only accumulations of a given size A_0 or greater. Although the choice of

A_0 is arbitrary to a degree, A_0 might be chosen small enough to encompass all pools detectable by use of current technology. It *must* be chosen small enough to include pools producible at a price far greater than that currently obtaining, so as to avoid a confounding of the definition of subpopulation elements with price. A modified version of I^a thus suggested is:

> I^b The size distribution (in barrels or Mcf) of petroleum deposits in pools within a subpopulation is *truncated* lognormal with truncation point $A_0 > 0$.

When A_0 is chosen to be very small, I^a and I^b lead to essentially similar results, although the problem of inference about parameters of the underlying size distribution is somewhat more complicated given I^b. An interpretative advantage of I^b is that one does not have to rationalize the proposition that, when $A_0 = 0$, there are "pools" within the subpopulation being sampled which are so small as to have virtually zero probability of ever being discovered. In addition, the parameter N takes on added meaning for, given I^b, it denotes the number of elements in a set, each member of which can, in principle, be discovered and identified using current technology. Henceforth we shall refer to "assumption I" and distinguish between I^a and I^b only where necessary.

The probabilistic behavior of amounts of oil (gas) in place discovered by each discovery well *in order of discovery* is completely determined by a conjunction of assumptions I and II. That is, assumption II implies that, "on the average," the larger (in size of oil [gas] in place) pools will be found first and, as the discovery process depletes the number of undiscovered pools in a subpopulation, discovery sizes will (again, "on the average") decline.

Our third assumption structures the behavior of the success ratio within a play once it has begun. Often a play begins with a stroke of geologic insight. With this insight, application of geophysical technology coupled with geologic analysis will identify a population of prospects, some of which will be pools and others of which will be dry. Letting S denote the sum of sedimentary volumes of all undiscovered pools in the play, and letting U denote the sum of sedimentary volumes of all undrilled prospects that are potentially identifiable by use of currently available exploration technology, we state:

> III^a The probability that an exploratory well will discover a new pool is equal to
> $$\frac{\kappa S}{\kappa S + U},$$
> where $\kappa > 0$ is a constant.

The constant κ is to be interpreted as an index of the difficulty (or ease) of discovery of pools within a given subpopulation *once a play has started within it*. Hence it may vary among subpopulations. For example, lense-type stratigraphic traps are more difficult to identify by seismic means than are pinnacle reefs, and thus might be assigned a smaller value of κ.

Assumption III is an extension of the idea behind assumption II (sampling proportional to size) and says that the probability of a discovery of *any* size shares the same general property. If the value of κ—the index of difficulty (or ease) of discovery—is 1, then drilling is "random" in the sense that predrilling exploration technology does not enhance the probability of discovery. Exploratory drilling in this particular case is like throwing darts into a three-dimensional volume, where each piece of equal volume has the same probability of being hit no matter where it is located. Even in this special case, the probability of discovery will change as $\kappa S/\kappa S + U$ changes with each exploratory well drilled.

A variant of III[a] is to replace "volume" by "areal extent"; we call this assumption III[b]. It is a more natural assumption than III[a] in certain respects. If drilling is completely random with respect to longitude and latitude, then the probability that a generic pool will be discovered is exactly equal to the ratio of the areal extent of the pool to the total areal extent of all undrilled prospects in that pool's geologic producing zone, and III[b], not III[a], is the relevant assumption. In fact, the areal extent and volume of many (but not all) pool types are highly correlated. Where such is the case, III[a] and III[b] are, in the use to which we shall put them, almost interchangeable.

Assumptions I, II, and III imply that once a play has begun the probability of discovery decreases on the average as the play unfolds. Although descriptively harsh—there are plays in which the success ratio continues to rise for a time after drilling of the initial discovery well—the specific functional form for the probability of success within a play implied by assumptions I, II, and III is fairly simple and allows us to calculate an estimate of κ.

Assumptions I, II, and III describe the physical evolution of a play, once it has begun. To articulate accurately in mathematical terms how and when a play begins is substantially more difficult. Geologic knowledge generated by seismic, gravity, magnetic, and surface surveys and analysis of exploratory well data, costs, and prices are determinants of the probability that a new play will begin at a given point on either a time scale or a scale composed of cumulative exploratory wells drilled. The spatial configuration and geographic location of sediments also play an important role. There are, nevertheless, several simple, descriptive assertions about the genesis of a play that lead to plausible postulates about the occurrence of a new play at a given point on a scale of cumulative number of wells drilled.

1. The cumulative number of exploratory wells drilled in the province is an index of geologic knowledge.
2. As the volume of unexplored sediment in the province decreases, so does the likelihood that a new play will occur.
3. Exploratory wells drilled in an existing play (*intensive* wells) are less likely to lead to a new play than wells drilled in an area not contiguous to an existing play (*extensive* wells).

The cumulative number of exploratory wells drilled is at best a crude surrogate for geologic knowledge. However, geologic knowledge does grow as the number of wells drilled grows, and so the latter is an index of the degree to which the geology of the region is understood. Assertion 3 suggests that the interarrival times between successive plays, measured on a scale of exploratory wells drilled, on the average becomes shorter as the proportion of extensive wells per well drilled becomes larger. Assumption IV articulates this idea more carefully, although considerable further refinement of it is necessary before it can be used to structure a probabilistic model of interarrival times between successive plays.

IV   Interarrival times between successive plays are uncertain quantities. The mean time between two successive plays, measured on a scale of cumulative exploratory wells drilled, (a) increases with an increase in the proportion of wells drilled extensively subsequent to the beginning of the first of these two plays, and (b) increases as the volume of unexplored sediment in the province decreases.

The analog of assumption IV for interarrival times measured on a time scale requires consideration of costs, prices, and investor behavior in the face of uncertainty (i.e., the economic returns to exploratory ventures within each subpopulation).

BACKGROUND FOR ASSUMPTIONS I, II, III, AND IV

A variety of studies supports the assertion that the size distribution of oil (gas) pools is adequately represented by a lognormal distribution.

Allais (1957), in a large-scale study of mineral resources of the Sahara Desert, concluded that the lognormal distribution provided a surprisingly good fit to frequency histograms of the *value* of deposits of ores such as iron, copper, gold, zinc, diamonds, etc. Krige's (1951) analysis of gold deposits in the Witwatersrand was, perhaps, the first to use the lognormal distribution as a characterization of the size distribution of a mineral deposit, and Allais' study strongly reinforced the reasonableness of the law in this context. Oil and gas were notably absent from the list of mineral resources treated in the published version of Allais' paper (an expected omission in view of the political complexities of France's relations with Algeria and her desire to hold onto the vast mineral resources of the region, irrespective of whether Algeria became independent).

Arps and Roberts' (1958) study of Cretaceous fields on the eastern flank of the Denver-Julesburg basin (using a large sample and grouping data) lent credence to the hypothesis that the size distribution of petroleum deposits is lognormal. Although Arps and Roberts proceeded heuristically, eschewing standard statistical testing procedures, their data plot very close to a straight line, even in the extreme right tail, when plotted on lognormal probability paper (cf Kaufman, 1963).

Kaufman (1963) examined what can be regarded at best as incomplete data and found that Arps and Roberts' assumption of lognormality was not unreasonable. McCrossan (1969), using the Alberta Province Energy Resources Conservation Commission's detailed compilation of data on individual pools discovered in Alberta, did a similar but more refined analysis. He first classified pools according to geologic type and then plotted fractile estimates derived from each sample so derived on lognormal-probability paper. Within classes composed of 50 or more pools (e.g., Viking Reef), the lognormal distribution provided a good visual fit. He also showed that one plausible explanation of the appearance of bimodality and/or deviations from lognormality in the tails is that observations from geologically distinct populations are being mixed together.

The studies by Allais (1957), Kaufman (1963), and McCrossan (1969) considered sizes in order of observation as observing values of a sequence of mutually independent and identically distributed random variables. In fact, oilmen long have observed that, within a play, the larger pools tend to be found first and the average size of new discoveries decreases as the play matures. Thus the process of observing pool sizes in order of discovery is more akin to sampling without replacement and proportional to (random) size than to sampling values of independent, identically distributed random variables.

The conjunction of these two features introduces novel complications and renders more difficult a careful theory of inference about parameters of the underlying size distribution—and about what remains to be discovered. If a play regarded as a sampling process possesses both of these attributes, there is a possibility of serious error in making inferences under the assumption that they are not present, and care should be taken to determine, in a systematic way, under what conditions these attributes must be explicitly taken into account.

More specifically, it is reasonable to postulate that discovery sizes in order of observation are generated by sampling without replacement from a *finite population* of pools whose sizes (area, volume) are generated by yet another random process; the finite population of pools is a random sample from a hypothetical infinite population (a *superpopulation*) whose size distribution is of known functional form. This characterization of sampling from a finite population is well known in the statistical litera- ture and has been used to develop classical, fiducial, and Bayesian procedures for estimation of finite population parameters (cf Cochran, 1939; Fisher, 1956; Ericson, 1969; and Palit and Guttman, 1973). However, if the sample drawn from the finite population is random, without replace- ment *and* proportional to size in the sense that the **pro**bability of observing the ith finite population element at the jth sample observation is equal to the ratio of the size A_i of that element to the sum of sizes of the as-yet-unobserved finite population elements, then, although the general framework is relevant, none of the specific techniques developed in the literature just cited can be applied directly. An additional complication is that the number of elements in the finite population is generally not known with certainty in this particular problem.

Uhler and Bradley (1970) analyzed the spatial occurrence of petroleum pools in Alberta, hypothesizing that the number of pools per unit area is describable by a negative binomial probability law. They obtained an excellent fit to actual frequencies of *observed* occurrences in "well- explored" areas (i.e., to the frequencies of pools per unit area dis- covered up to 1970). Their method provides one way of estimating the frequency of occurrence of pools in a given area. Drew (1972) conducted an empirical study of the spatial distribution of petroleum within land tracts in Kansas. Accounting the effect of land ownership on the number of deposits *discovered* per unit area in Kansas leads to a probability law substantially different from the negative binomial.

Cox (1969) discussed sampling proportional to size from an infinite population with particular reference to the sampling of textile fibres, and his results have relevance as the size of the finite population approaches infinity.

Arps and Roberts' (1958) model, Kaufman's (1965) recharacterization of it in terms of a system of differential equations, and Crabbe's (1969) modification of Kaufman's work obliquely embody the notion of sampling proportional to random size. However, the models in these three papers are formulated in such a way that rigorous statistical testing of the basic assumptions underlying them is difficult if not impossible. Drew (1974) conducted an interesting retrospective simulation study of explora- tion in the Powder River basin by using historical data on pool sizes and their location as an empirical base. This study descriptively embodies sampling proportional to size and without replacement. Drew carefully pointed out that, since his study is retrospective in structure, explicit predictions about future discoveries can be made only by assuming similarity between the resource base in the control area on which the simulation is conducted and an unexplored target area.

Assumption III asserts that the probability that an (intensive) exploratory well will discover a new pool within a given subpopulation is proportional to the ratio of the volume of undiscovered pools to the volume of potentially hydrocarbon-bearing sediment of that population's geologic type; it is a logical extension of assumption II. Although the assumption is plausible in nature and has appeared in disguised form in several papers, it has never been validated empirically. Ryan (1973a), in an important paper on the crude-oil discovery rate in Alberta, based

his analysis on a set of assumptions similar to II and III, the most
important one being an amalgam of deterministic versions of those assumptions: "The rate of discovery of oil in a play is proportional to the
undiscovered oil in the play and the knowledge of existence of the play."
A probabilistic version of this postulate, not empirically validated,
appears in Kaufman and Bradley (1972).

Ryan (1973a) was the first to investigate a deterministic model of
discovery consisting of a superposition of models of individual plays
on a scale of cumulative number of wells drilled. The differential
equation he obtained for additions to reserves per well drilled is a
deterministic version of the stochastic model that follows from assumptions I, II, III, and IV. Although extremely useful in providing rough
estimates of additions to oil in place from plays already known to be
in existence, his model contains no mechanism that generates new plays
as more wells are drilled. Ryan gives a thorough discussion of the
strengths and weaknesses of his model. In a second paper, Ryan (1973b)
point-forecasts growth of potential crude-oil reserves in Alberta as a
function of new-field wildcats drilled.

PROBABILISTIC MODEL OF AN INDIVIDUAL PLAY

Our mathematical description will be done backwards; that is, we
shall begin by describing in detail a submodel of amounts discovered
within a play per discovery in order of observation. The size of a discovery will be measured in stock-tank barrels of oil (Mcf equivalent if
gas) in place. We shall assume that assumptions I (lognormality of size
distribution) and II (sampling without replacement and proportional to
random size) hold. These two assumptions are sufficiently rich to guarantee logical completeness of this submodel. This submodel describes
both the outcomes of drilling (discovery or dry hole) and the size of a
discovery when one is made. It evolves on a scale of wells drilled in
a given play, and its properties are not influenced by either the rate at
which wells are drilled or by economic variables. A superposition of
individual plays on a time scale requires additional assumptions about
the influence of economic variables on the amount of exploratory effort
in a given time interval and how it is allocated among plays. In a later
report we shall develop a model of returns to drilling in each play
using assumption III.

EMPIRICAL SIZE DISTRIBUTIONS

A cursory examination of empirical size distributions of oil pools
shows that these distributions are usually unimodal and highly skewed,
having very long right tails; that is, a small proportion of observed
values is very large and a large proportion is very small. There is
an infinity of functional forms for unimodal distribution functions concentrated on zero to infinity and having long right tails. The problem
of deciding which particular functional form best fits observed data on
sizes of petroleum pools is quite complicated.

Some of the reasons are: (1) *reported* oil (gas) in place is only
an engineering estimate, not a direct observation of *actual* oil (gas)
in place; and (2) the accuracy of reported oil (gas) in place in a
pool generally will improve as the pool is developed, so that the
initial estimate may be far from that given when the pool is nearly
depleted. In addition to reporting bias, *truncation* of sample observa-

tions may be present. (3) An exploratory well may yield only a "show" of oil (gas). In such cases hydrocarbons are present, but in an amount far below the economic breakeven point. Such "pools" will not be exploited and the well may be reported as a dry hole. Another complicating feature is that, when sampling is without replacement and proportional to random size, the distribution of sample observations is not the same as the size distribution of pools in nature.

One can buttress a choice among functional forms by appealing to basic principles or postulates describing the process of hydrocarbon deposition. One popular postulate (cf Matheron, 1955; Rodionov, 1964) is that the accessory mineral content of rock is generated by a law of proportionate effect. This leads (via a central limit theorem) to the assertion that mineral deposits in rock are approximately lognormal. Mandelbrot (1960) has argued that *stable* probability laws are a more appropriate representation. Though having the virtue of skewness and very fat right tails, stable densities (concentrated on $[0,\infty]$) are analytically intractable and may be without mean or variance. The only analytically tractable version is, for $\Theta > 0$,

$$f_S(Y|\Theta) = \begin{cases} \dfrac{\Theta}{\sqrt{2\pi}}\, Y^{-3/2} e^{-\frac{1}{2}\Theta^2/Y} & \text{for } Y > 0; \\ \\ 0 & \text{otherwise.} \end{cases} \tag{1}$$

This density has neither mean nor variance. It can be labeled "inverted gama," since $X = 1/Y$ has a gamma density.

Unfortunately, it is possible to distinguish with precision between lognormal and stable probability laws only with very large sample sizes. Prokhorov (1964) gave an instructive account of why this is true. His mathematical development is motivated by consideration of the absolute mineral content in a given volume of rock, but it is relevant in general. He stated that, "It is possible to distinguish the exponential distribution with density $\exp\{-Y\}$, $Y > 0$ from a lognormal distribution with density $(1/\sqrt{2\pi}Y)\exp\{-\frac{1}{2}(\log Y)^2\}$, $Y > 0$ with errors of Type 1 and Type II equal to .05 only on the basis of a sample size close to 100." Prokhorov's assertion is based on a chi-squared test applied to grouped data.

In the present case, an attempt to distinguish between competing hypotheses about the size distribution of deposits is even more difficult if sampling is without replacement and proportional to size, for the sampling distribution for observed sizes is not the same as the distribution that generates the size of pools deposed by nature. A crude pretest is to test the hypothesis that *observed* pool sizes are lognormally distributed against the specific hypothesis that they are gamma distributed.

The next section presents the results of such a test, one designed so that both hypotheses may be simultaneously rejected or accepted. One would expect that both hypotheses will be accepted if the sample size is small, and both hypotheses will be rejected if the sample size is large and the size distribution in nature is lognormal. At the 1-percent level of significance, this latter event occurs only once among 21 samples—but, significantly, the sample size for this case is very large by comparison with all other cases. When plotted on lognormal probability paper (cf following section), most of the 21 samples show substantial deviation from lognormality in the extreme tails. The implications are that (1) much larger sample sizes may result in decisive rejection of both the lognormal and gamma hypotheses, and (2) a better test is lognormality against the specific hypothesis that the probability law of observed sizes has a functional form dictated by assumptions I and II.

LOGNORMAL VERSUS GAMMA

Karen Sharp[5] has developed a program for testing the hypothesis of lognormality against the (specific) alternative hypothesis that the underlying probability law is gamma. Her test procedure is based on methods developed by Jackson (1969) and Cox (1962), methods that seem to be more discriminating than a chi-squared test of a specific hypothesis against an unspecified alternative.

She says:

The test has two distinct but equally necessary parts. They can be described basically as follows. Four statistics are calculated from the sample distribution. If the distribution is actually lognormal these four estimated parameters will bear a certain relation to each other. The first half of the test is constructed to measure how closely the actual numbers conform to this relationship. If they do not conform closely the assumption of a lognormal distribution is rejected. The two way test is symmetrical. If the distribution is actually gamma, the four parameters will be related in another specific way which has been derived from the nature of the two distributions. A test statistic compares the actual numbers to this assumed relation. This two way test is necessary to avoid acceptance of a false alternative. It is possible that the frequency distribution observed conforms poorly to neither of the two distributions tested; in this case each test will signal rejection of the corresponding assumption. It is also possible that the sample does not provide enough information to permit a choice between the distributions; neither test will reject the hypotheses.

Each distribution is specified by two parameters. The four "statistics" calculated are the sample parameter estimates. Each half of the test uses the standard method of testing a null hypothesis against a single possible alternative.* With one half of the test, a lognormal distribution with parameter (μ, σ^2) is assumed. Maximum likelihood estimates, $\hat{\mu}$ and $\hat{\sigma}^2$, of the parameters are calculated from the sample data. Knowledge of the form of the two distributions permits the derivation of shadow gamma parameters as functions of $\hat{\mu}$ and $\hat{\sigma}^2$. These shadow parameters $\beta_1(\hat{\mu}, \hat{\sigma}^2)$ and $\beta_2(\hat{\mu}, \hat{\sigma}^2)$ are estimates of the numbers that one would get from attempting to fit a gamma distribution to a sample which is actually lognormal with parameters $\hat{\mu}$ and $\hat{\sigma}^2$. If the hypothesis of a lognormal distribution is true, then a maximum likelihood estimate $\hat{\beta}_1$ and $\hat{\beta}_2$ of gamma parameters β_1 and β_2 will closely approximate $\beta_1(\hat{\mu}, \hat{\sigma}^2)$ and $\beta_2(\hat{\mu}, \hat{\sigma}^2)$. The larger the sample size the closer these estimates should be to $\beta_1(\hat{\mu}, \hat{\sigma}^2)$ and $\beta_2(\hat{\mu}, \hat{\sigma}^2)$. The test statistic measures the difference between $(\hat{\beta}_1, \hat{\beta}_2)$ and $(\beta_2(\hat{\mu}, \hat{\sigma}^2), \beta_2(\hat{\mu}, \hat{\sigma}^2))$ and each half of the test has a standard normal distribution. If the hypothesis of a lognormal distribution is true the test statistic will be approximately zero, and hence insignificant. If the test is significantly different from zero (by comparison to the usual table of normal deviates) the hypothesis can be rejected. The other half of the test is similar, involving shadow parameters $\mu(\hat{\beta}_1, \hat{\beta}_2)$ and $\sigma^2(\hat{\beta}_1, \hat{\beta}_2)$.

*In a simple example with a null hypothesis H_0: $\bar{x} = a$, a simple alternative would be H_1: $\bar{x} = b$ as compared to a composite hypothesis such as H_2: $\bar{x} > a$.

Table 1 displays the results of applying Sharp's test to 21 categories of pools in Alberta. The three columns under "Lognormal Assumed" display

[5]Economics Department, Energy Resources Conservation Board.

TABLE 1.   LOGNORMALITY VERSUS GAMMA HYPOTHESIS TESTS

Formation Name	File Data Code No.	No. of Pools	Lognormal Assumed			Gamma Assumed		
			$\hat{\mu}$	$\hat{\sigma}^2$	Test Stat. 1	$\hat{\beta}_1$	$\hat{\beta}_2$	Test Stat. 2
Belly River	1260	38	8.40950680	2.10364723	0.263183534	13849.3125	0.554931641	-3.03674507
Cardium	1760	44	8.37104893	5.10227776	0.595982373	203769.813	0.192154884	-6.39682293
Viking	2180	33	7.89568520	3.73327637	0.241688788	22155.7852	0.322883606	-3.73119068
Blairmore	2440	49	7.84102917	1.94996071	1.41248798	9475.24219	0.485148430	-4.75621128
Mannville	2480	48	8.08687782	2.77233028	-0.021886862	12373.6016	0.478538513	-3.1157846
Upper Manville	2500	26	8.21560574	2.25119114	-0.371570587	9978.19141	0.619585037	-1.65595150
Manville	2760	29	8.64087200	2.02218151	-0.550133765	13381.2734	0.701862335	-1.44166660
Glauconitic	3000	18	7.58537674	2.32115936	0.507498503	8215.33203	0.451807976	-2.66753006
Lower Manville	3100	40	7.86137009	1.25508881	1.44930267	5788.84766	0.746592522	-3.72580051
Basal Manville	3200	21	7.49146748	1.56010723	0.545270801	4436.37891	0.670944214	-2.39769173
Basal Quartz	3340	21	7.45048618	2.06439209	0.768393576	6577.37891	0.477160454	-3.00600052
Jurassic	4460	16	8.70669174	2.46884727	-0.333106518	17360.5625	0.587364197	-1.25730038
Rundle	6100	20	9.08085537	5.37532043	-0.49826229	83624.0625	0.305179596	-2.15662575
Pekisko	6420	32	7.93286037	2.11803722	-0.310268044	7340.71875	0.633093834	-1.96602249
D-2	6960	53	8.56196022	2.71537876	0.386777341	23724.3359	0.430091858	-4.16236115
D-3	7200	68	8.90444374	4.00139904	0.201316953	64098.0273	0.316000938	-5.17798424
Beaverhill Lake	7440	19	10.8841162	5.12460613	-0.408864856	268253.125	0.406288147	-0.922035396
Muskeg	7820	52	6.94740105	0.747285724	-0.0768762231	1507.11523	1.49411349	-1.99720287
Keg River	7880	195	7.93582439	1.54143810	3.76589584 ¤	8191.00391	0.577983856	-9.50142288
Keg River	7881	109	7.46872520	0.641348362	1.82708645	2519.18335	1.52278233	-4.55488873
Keg River	7882	31	7.07624340	0.890673816	-2.84538555 ¤	1571.19336	1.91448784	2.09837818

for each category an estimate $\hat{\mu}$ of the mean μ of the (natural) logarithm of reservoir size, an estimate $\hat{\sigma}^2$ of the variance σ^2 of the logarithm of pool size (in units of 1,000 STB), and a test statistic. These three numbers are computed under the hypothesis that the sample of observed pool sizes in each category is generated by a lognormal process. Similarly, the three columns under "Gamma Assumed" display for each category estimates $\hat{\beta}_1$ and $\hat{\beta}_2$ of parameters β_1 and β_2 of a gamma density with mean β_1 and variance β_1^2/β_2 fitted to observations in each category, and a test statistic, under the hypothesis that these observations are generated by a gamma process.

The decision rule for a 1-percent (two-tailed) test of significance is as shown by Table 2. For example, where $|T_1| > 2.576$ and $|T_2| > 2.576$, the conditional probability of making an error in asserting that the observations leading to T_1 and to T_2 come from neither a lognormal nor a gamma process is less than, or equal to, 0.01.

Table 3 is a summary of test results. At the 5-percent level of significance, *in no case* is the gamma hypothesis accepted and the lognormal hypothesis rejected. In the four cases where both are accepted, sample sizes are relatively small, ranging from 16 to 29. Both hypotheses are rejected for Keg River 7880 (195 observations) and Keg River 7882 (31 observations). In 15 of 21 cases the lognormal hypothesis is accepted and the gamma hypothesis rejected.

At the 1-percent level of significance, the gamma hypothesis is accepted and the lognormal hypothesis rejected in one instance—Keg River 7882. In no case are both hypotheses rejected. Here also the lognormal hypothesis is favored, although less strongly than at the 5-percent level of significance.

Although this battery of tests makes a prima facie case for favoring the lognormal hypothesis over the gamma hypothesis, it is dangerous to conclude that the process by which new pools are discovered (the observational process) is lognormal. In particular, the test procedure used here is not terribly sensitive to tail behavior until sample size becomes very large. We plotted a number of samples on lognormal probability paper (Fig. 1); it is clear that the empirical cumulative distribution functions have, on the average, much fatter right tails than one would expect if the observations were in fact values of mutually independent, identically distributed, lognormal random variables. As we shall show, approximate lognormality in the interquartile range with a fatter-than-lognormal right tail is precisely what is implied by assumptions I and II taken together.

LOGNORMAL VERSUS PROBABILITY LAW IMPLIED
BY ASSUMPTIONS I AND II

Testing the hypothesis of lognormality of observed pool sizes against the hypothesis that pool sizes are gamma distributed is informative but logically out of joint with our model of the discovery process, since assumptions I and II taken together lead to the hypothesis that neither hypothesis is appropriate. It is informative in that we can conclude with reasonable certainty that a gamma probability law is not as accurate a characterization of the sampling density of observed sizes as a lognormal probability law—although, as we shall show, the latter is, in turn, a less satisfactory hypothesis than one dictated by assumptions I and II (cf following section). If we call the lognormal hypothesis "H_1" and the hypothesis that the sampling density is as displayed in the following section, "H_2," a Bayesian test of hypothesis H_1 versus H_2 is

TABLE 2.   DECISION RULES FOR 1-PERCENT TEST OF SIGNIFICANCE*

Test Stat. 1 $\equiv T_1$	Test Stat. 2 $\equiv T_2$	Conclusion
$\lvert T_1 \rvert$ > 2.576	$\lvert T_2 \rvert$ > 2.576	Reject both distributions
> 2.576	< 2.576	Gamma distribution is more appropriate
< 2.576	> 2.576	Lognormal distribution is more appropriate
< 2.576	< 2.576	Not enough information; either distribution could be used

*For a 5-percent level of significance, replace 2.576 with 1.96.

possible; asserting that a priori H_1 and H_2 are equally likely, we compute the odds, posterior to observing the data, that the data were generated according to H_2 rather than H_1.[6]

DEFINITION OF SAMPLING PROCESS FOR DISCOVERY SIZES

Basically, assumption II is that discovery sizes in order of observation are generated by sampling without replacement from a finite population of pools whose sizes constitute a sample from a hypothetical infinite population. We shall call this latter population a *superpopulation*. Observations of discovery sizes come about as follows: nature generates a sequence of values $A_1,\ldots,A_N$ of N mutually independent, identically distributed random variables (rvs) $\tilde{A}_1,\ldots,\tilde{A}_N$ with common density f concentrated on $[0,\infty)$; f characterizes the superpopulation from which the sizes $A_1,\ldots,A_N$ of N pools deposed by nature are drawn. These values are *not* observed in the order in which they are generated. Rather, elements of the finite set $Q_N = \{A_1,\ldots,A_N\}$ of pool sizes is sampled without replacement and proportional to size; the finite population being sampled is Q_N.

Where Q_N is known, assumption III says that the probability of observing $A_{i_1}, A_{i_2},\ldots,A_{i_n}$, $n \leq N$, in that order, is (upon relabeling elements of Q_N so that $[i_1,i_2,\ldots,i_n] = [1,2,\ldots,n]$):

$$P\{(1,2,\ldots,n)\,|\,Q_N\} = \prod_{j=1}^{n} A_j/(A_j+\ldots+A_N). \tag{2}$$

Clearly, the values $A_1,\ldots,A_N$ are *not* known a priori, and, even after observing values of n of the A_i's, $N-n$, such values remain unknown with certainty.

Let Y_j denote the observed value of the jth pool discovered, define $\underline{Y} = (Y_1,\ldots,Y_n)$ as the vector of observations in a sample of size $n \leq N$, and assume that f is a member of a class of densities (all of whose members are concentrated on $[0,\infty]$) indexed by a parameter $\underline{\theta} \in \Theta$ so that $\tilde{A}_i$

[6]See, for example, Chapter 2 of Zellner (1970).

TABLE 3.   RESULTS OF LOGNORMALITY VERSUS GAMMA HYPOTHESIS TESTS

At 5-Percent Level of Significance

		Lognormal Hypothesis		
		Accept	Reject	
Gamma Hypothesis	Accept	4	0	5
	Reject	15	2*	16
		19	2	21

At 1-Percent Level of Significance

		Lognormal Hypothesis		
		Accept	Reject	
Gamma Hypothesis	Accept	8	1**	9
	Reject	12	0	12
		19	2	21

*Keg River 7880 and Keg River 7882.
**Keg River 7882.

has density $f(\cdot|\underline{\theta})$. Then, if $\underline{\theta}$, N, and infinitesimal intervals $dY_1,\ldots,$ dY_n are known, the probability of observing $\tilde{Y}_1 \in dY_1,\ldots, \tilde{Y}_n \in dY_n$ in that order (or, equivalently, of observing $\underline{\tilde{Y}} \in d\underline{Y}$) is:

$$P\{\underline{\tilde{Y}} \epsilon d\underline{Y}|\underline{\theta},N\} = [(N)_n \prod_{j=1}^{n} Y_j f(Y_j|\underline{\theta})dY_j] \int_0^\infty f^{*N-n}(S|\underline{\theta}) \prod_{j=1}^{n} (Y_j+\ldots+Y_n+S)^{-1}dS. \quad (3)$$

Here, $(N)_n \overset{\text{def}}{=} N(N-1)\ldots(N-n+1)$ and f^{*N-n} is the density of the sum of $N-n$ $\tilde{A}_i$'s. (This expression for $P\{\underline{\tilde{Y}} \epsilon d\underline{Y}|\underline{\theta},N\}$ is predicated on the existence of the integral in equation 3.)

Using equation 3, we may, in principle, make inferences about the parameter $\underline{\theta}$ when it is not known with certainty (as is usually the case), as well as about the sum of unobserved finite population elements. This sum is of particular interest because it constitutes the sum total of undiscovered oil (gas) in place within the play being sampled.

PROPERTIES OF ASSUMPTIONS I AND II VIA MONTE CARLO SIMULATION

In order to give an intuitive "feel" for the implications of assumptions I and II, we describe here the output of a Monte Carlo simulation of the sampling process for discovery sizes dictated by these assumptions. Our attention is focused on three objects:

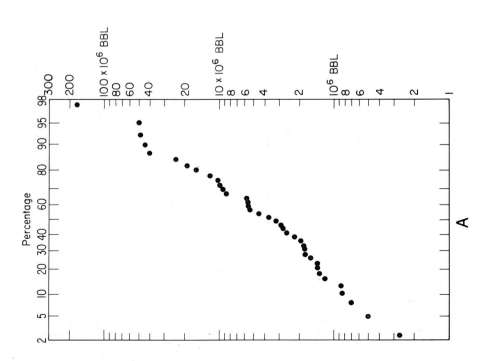

FIG. 1—Samples plotted on lognormal probability paper. **A.** Belly River, 38 observations; **B.** Lower Manville, 40 observations; **C.** Keg River 7880, 195 observations; **D.** Manville, 48 observations; **E.** D-2, 53 observations. (Continued on next two pages.)

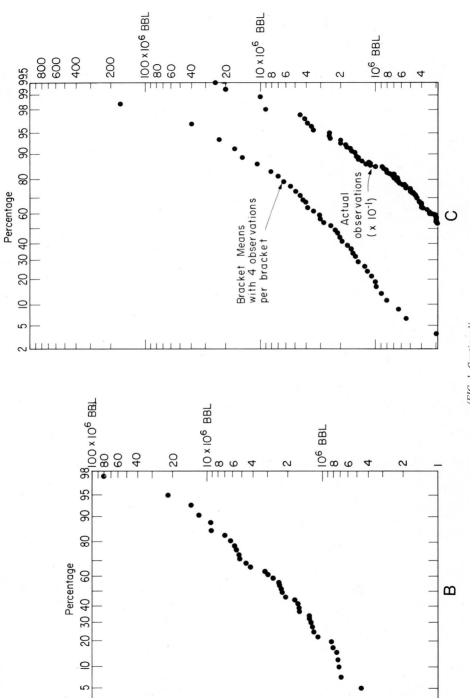

(FIG. 1. Continued)

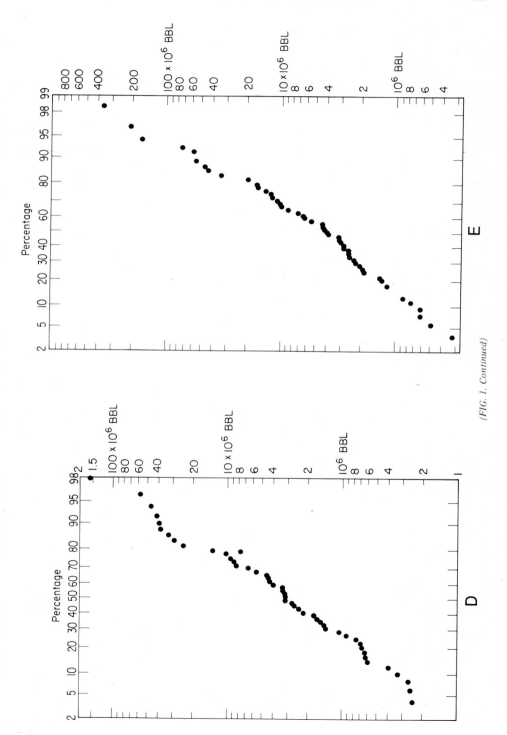

(FIG. 1. Continued)

1. The probability distribution $P\{\tilde{\underline{Y}} \epsilon d\underline{Y} | \theta, N\}$ of observed sizes;
2. The probability distribution of undiscovered sizes where $\tilde{\underline{Y}} = \underline{Y}$;
3. The probability distribution of the mean $\tilde{\tilde{S}}_{N-n}$ of undiscovered sizes where $\tilde{\underline{Y}} = \underline{Y}$.

We will examine (1) and (2) relative to assumption I in the following way: assume that the size distribution of petroleum deposits in pools is lognormal with parameter (μ, σ^2); that is, $\tilde{A}_1, \ldots, \tilde{A}_N$ are mutually independent with common density for $-\infty < \mu < +\infty$ and $\sigma^2 > 0$:

$$f_L(A | \mu, \sigma^2) = \begin{cases} \dfrac{1}{\sigma\sqrt{2\pi}} e^{-\frac{1}{2}(\log_e A - \mu)^2/\sigma^2} \dfrac{1}{A} & \text{if } A > 0; \\ \\ 0 & \text{otherwise.} \end{cases} \qquad (4)$$

To simulate observations generated according to assumptions I and II, we first generate values $A_1^{(i)}, \ldots, A_N^{(i)}$ according to equation 4, and then, given $\{A_1^{(i)}, \ldots, A^{(i)}\} \equiv Q_N^{(i)}$, we generate $\underline{Y}^{(i)}$ according to equation 1. Here the index i indexes replications of our Monte Carlo experiment. It is obvious (see equation 2) that elements of $\tilde{\underline{Y}}^{(i)}$ are neither independent nor marginally, identically distributed as lognormal. However, suppose that we incorrectly assume that they are and examine fractile plots of the $Y_j^{(i)}$'s on lognormal probability paper as if they constitute independent sample observations from a lognormal process—as has been done by several authors. How does the empirical cumulative function so generated deviate from lognormality? Undiscovered sizes, the complement $U_N^{(i)}$ of $\{Y_1^{(i)}, \ldots, Y_n^{(i)}\}$ in $Q_N^{(i)}$, are treated similarly in our experiment.

The following are the salient facts:

1. On lognormal probability paper the graph of fractiles computed from $Y_1^{(i)}, \ldots, Y_n^{(i)}$ is, on the average, close to linear within the interquartile range but exhibits a fatter right tail than that of a lognormal distribution. The graph is tilted, having a smaller slope than that exhibited by the (straight line) graph of fractiles of the underlying lognormal distribution of the $\tilde{A}_j^{(i)}$'s, and lies entirely *above* the latter graph.

2. The graph of fractiles computed from undiscovered sizes is, on the average, linear within the interquartile range but exhibits a smaller right tail than that of a lognormal distribution. It is also tilted, having a smaller slope in the interquartile range than that of the graph of fractiles of the underlying lognormal distribution, and lies entirely *below* the latter graph.

3. The mean $E(\tilde{Y}_j | \theta, N)$ of the size of the jth discovery is far above the mean of the underlying (lognormal) population for small values of j, but it declines faster than exponentially with increasing j at first and then declines slower than exponentially.

In order to reduce the effects of Monte Carlo sampling variability, we replicated our experiment 1,000 times. For the following example, graphs of the sample means of fractile estimates cited above are shown in Figures 2-19. In addition, we computed sample estimates of the marginal means of the $\tilde{Y}_j$'s and of the covariance structure of $\underline{Y}$. Coincident with our intuition:

4. The distribution of $\tilde{Y}_j$ is very close to lognormal for small values of j but, as j increases, right-tail probabilities become smaller than those of a lognormal distribution with the same mean and variance.

The graphs displayed in Figures 2-16 were generalized by averaging 1,000 Monte Carlo replications of sampling n pool sizes without replacement and proportional to random size from a finite population of N pools. Values chosen for n and N were:

$N = 1,200$ and $n = 10, 20, 30, 40, 50, 75, 100, 150, 200;$
$N = 600$ and $n = 10, 20, 30, 40, 50, 75, 100, 150;$
$N = 300$ and $n = 10, 20, 30, 40, 50, 75, 100;$
$N = 150$ and $n = 10, 20, 30, 40, 50, 75;$
$N = 100$ and $n = 10, 20, 30, 40, 50.$

Elements of the finite population have sizes generated according to a lognormal probability law with parameters $\mu = 6.00$ and $\sigma^2 = 3.00$. In 10^3 bbl of oil in place, the corresponding density has median $\exp\{6.00\}$ = 403.4 and mean $\exp\{\mu+\tfrac{1}{2}\sigma^2\}$ = 1,808.

Figures 2-11 display simulated versions of (1) the expectation of the empirical cumulative distribution function of observed pool sizes and (2) the expectation of the empirical cumulative function of sizes of pools remaining to be discovered, computed under the assumption that observed sizes are mutually independent and identically distributed—which they are not. The graph is plotted on lognormal probability paper with ordinate expressed in natural logarithms and abscissa in probabilities of less than the corresponding ordinate values. By computing this expectation and plotting it as described, we can see how far "off" we are by making the assumption that observed sizes are in fact independent and identically distributed as lognormal. The right tail curves noticeably away from a straight line and is displaced upward from the straight line (graph) of the underlying population's cumulative distribution function.

The graph of the expectation of the empirical cumulative distribution function for given N and n, plotted in a similar manner, lies below the straight line (graph) of the underlying population's cumulative distribution function, and the right tail becomes progressively "thinner" as a larger proportion of the finite population of pools is sampled.

Figures 12-16 display the graphs of the expectation of the empirical distribution function of the observed Y_j's plotted in a similar manner.

Figure 17 displays the graph of the (simulated) mean of observed pool sizes in order of discovery for several values of N. The first few discoveries have mean sizes which are orders of magnitude greater than the mean size—$\exp\{\mu+\tfrac{1}{2}\sigma^2\}$ = 1,808,000 bbl in place—of the underlying lognormal population. For example, when $N = 100$, the mean size of the first discovery is more than 12 times the underlying-population mean, and, when $N = 1,200$, it is more than 17 times this mean! The mean size of the jth discovery declines at a very rapid rate; for example, for $N = 100$, only the first 20 discoveries have mean sizes larger than the underlying-population mean.

Figure 18 shows graphs of (simulated) means of observed pool sizes in order of discovery for several values of N as a function of the proportion of undiscovered pools that have been discovered; $E(\tilde{Y}_j|\underline{\theta},N)$ is plotted as a function of $j/N-j$. As N varies from 100 to 1,200, the graphs remain virtually indistinguishable. If they are in fact indistinguishable, the implication is that $E(\tilde{Y}_j|\underline{\theta},N)$ is the same for every pair (j,N), $j < N$, of positive integers such that $j/N-j$ is a constant; that is, $\lambda = j/N-j$, $j=1,2,\ldots,N-1$ is the "natural" scale for $E(\tilde{Y}_j|\underline{\theta},N)$.

A least-squares fit of a third-degree polynomial to simulated values of $\log E(\tilde{Y}_j|\underline{\theta},N)$ of the form

$$\log E(\tilde{Y}_j|\underline{\theta},N) = \beta_0 + \beta_1\left(\frac{j}{N-j}\right) + \beta_2\left(\frac{j}{N-j}\right)^2 + \beta_3\left(\frac{j}{N-j}\right)^3 + \text{error} \qquad (5)$$

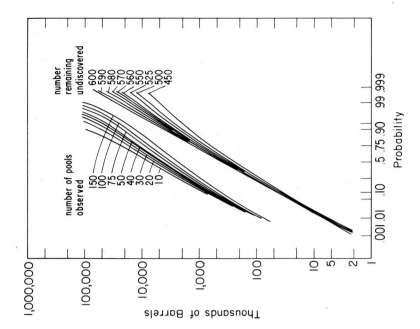

FIG. 3 – Simulated cumulative distribution functions for observed pool sizes and for pools remaining undiscovered when $N = 600$.

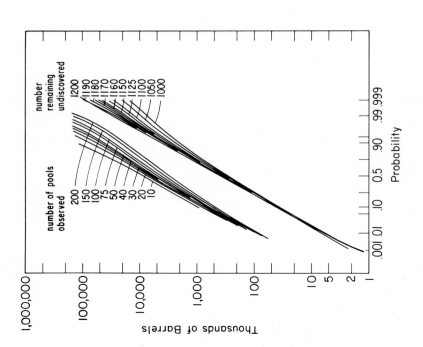

FIG. 2 – Simulated cumulative distribution functions for observed pool sizes and for pools remaining undiscovered when $N = 1,200$.

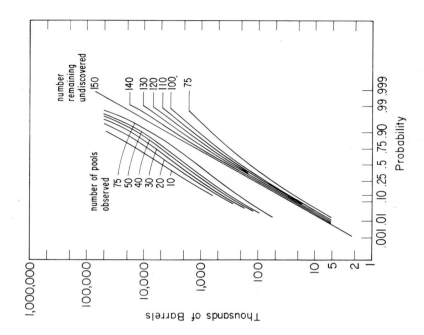

FIG. 5.—Simulated cumulative distribution functions for observed pool sizes and for pools remaining undiscovered when $N = 150$.

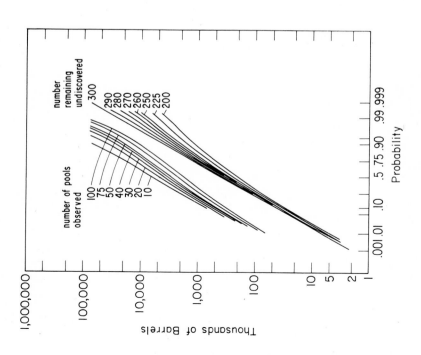

FIG. 4.—Simulated cumulative distribution functions for observed pool sizes and for pools remaining undiscovered when $N = 300$.

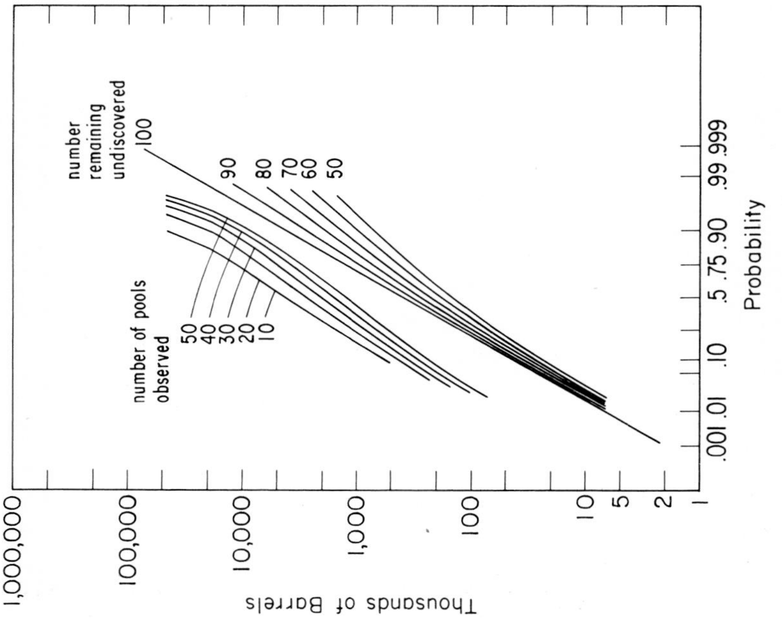

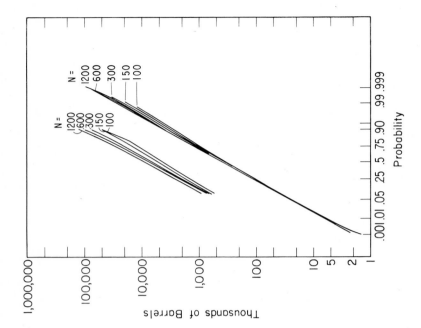

FIG. 7—Simulated cumulative distribution functions for observed pool sizes and for pools remaining undiscovered—for fixed sample size, $n = 10$.

FIG. 6—Simulated cumulative distribution functions for observed pool sizes and for pools remaining undiscovered when $N = 100$.

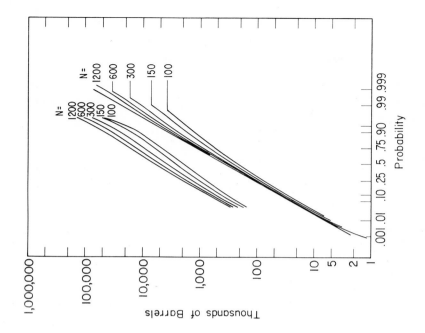

FIG. 9 – Simulated cumulative distribution functions for observed pool sizes and for pools remaining undiscovered – for fixed sample size, $n = 30$.

FIG. 8 – Simulated cumulative distribution functions for observed pool sizes and for pools remaining undiscovered – for fixed sample size, $n = 20$.

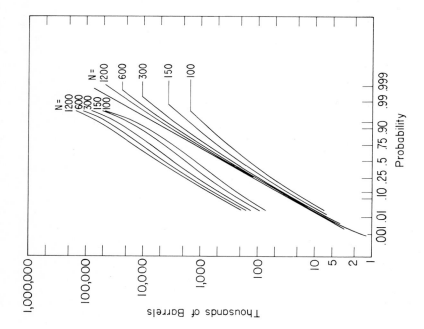

FIG. 11—Simulated cumulative distribution functions for observed pool sizes and for pools remaining undiscovered—for fixed sample size, $n = 50$.

FIG. 10—Simulated cumulative distribution functions for observed pool sizes and for pools remaining undiscovered—for fixed sample size, $n = 40$.

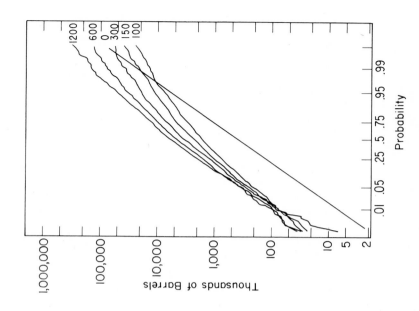

FIG. 13—Simulated cumulative distribution functions for size of 20th pool discovered when finite population size N = 100, 150, 300, 600, and 1,200.

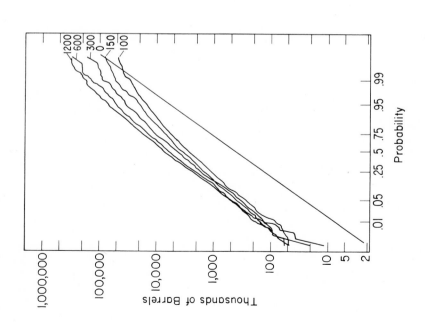

FIG. 12—Simulated cumulative distribution functions for size of 10th pool discovered when finite population size N = 100, 150, 300, 600, and 1,200.

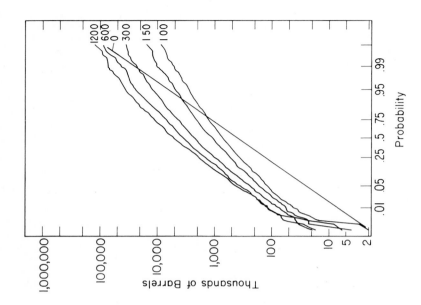

FIG. 15—Simulated cumulative distribution functions for size of 40th pool discovered when finite population size N = 100, 150, 300, 600, and 1,200.

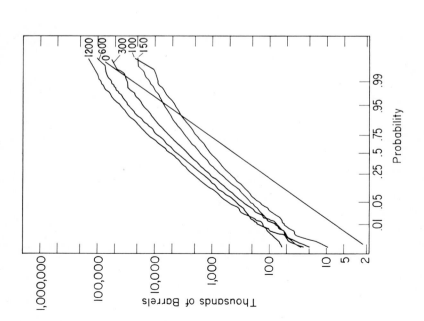

FIG. 14—Simulated cumulative distribution functions for size of 30th pool discovered when finite population size N = 100, 150, 300, 600, and 1,200.

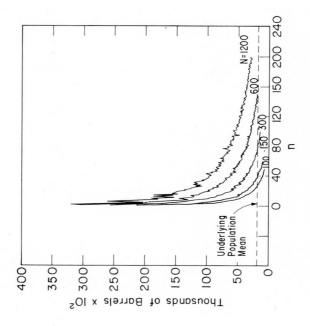

FIG. 17—Simulated means of size of nth pool discovered for finite population sizes $N = 100, 150, 300, 600,$ and $1,200$.

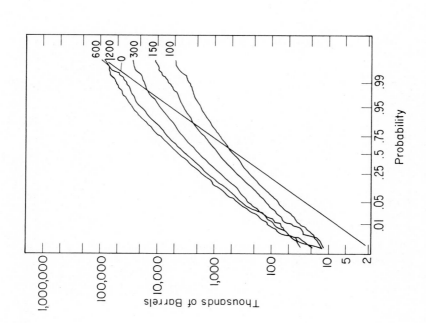

FIG. 16—Simulated cumulative distribution functions for size of 50th pool discovered when finite population size $N = 100, 150, 300, 600,$ and $1,200$.

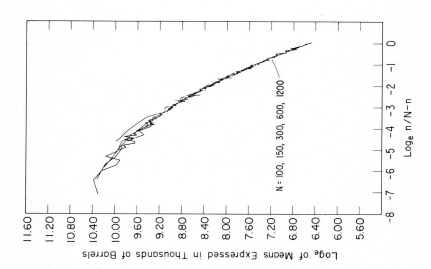

FIG. 19—Simulated means of size of nth pool discovered for finite population sizes N = 100, 150, 300, 600, and 1,200, displayed as a function of $\log_e n/N-n$, the proportion of undiscovered pools discovered.

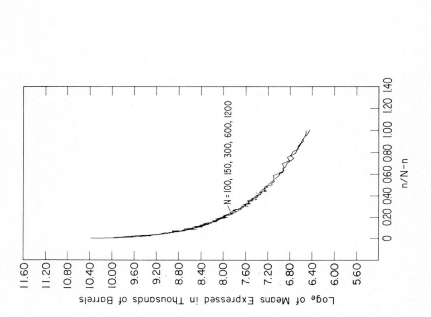

FIG. 18—Simulated means of size of nth pool discovered for finite population sizes N = 100, 150, 300, 600, and 1,200, displayed as a function of $n/N-n$, the proportion of undiscovered pools discovered.

TABLE 4. STANDARDIZED COEFFICIENTS

(N,n)	Constant Term	$j/N-j$	$(j/N-j)^2$	$(j/N-j)^3$	R^2
100,25	10.028	-3.349	4.517	-2.151	0.987
50	9.588	-3.336	4.507	-2.138	0.982
150,37	9.883	-3.047	3.779	-1.707	0.989
75	9.518	-3.238	4.226	-1.949	0.986
300,50	10.100	-3.154	4.014	-1.834	0.991
100	9.802	-3.260	4.259	-1.967	0.987
600,75	10.132	-3.054	3.827	-1.747	0.984
150	9.909	-3.153	1.064	-1.885	0.987

fits quite well. For the range of values of n and N shown in Table 4, *standardized* coefficients corresponding to the β_i's are reasonably stable.

PRELIMINARY CONCLUSIONS

The preliminary results reported here suggest that a model of discovery sizes based on assumptions I and II is a promising improvement over a variety of earlier models. In particular, a decline in the average size of discovery as the resource base is depleted appears as a logical consequence of these assumptions. However, before the model can be regarded as empirically valid, additional statistical testing of underlying assumptions and of the model's predictive accuracy must be done.

We believe that it is possible to interface our model with expert subjective judgment so as to generate probabilistic forecasts of discovery sizes for prospective plays in which no drilling has been done. That is, if one views our model as a generator of discovery sizes, one whose parameters are not known with certainty, and codifies expert judgment about these parameters in the form of subjective probabilities, the calculation of a predictive probability distribution for discovery sizes is conceptually straightforward, albeit computationally involved.

REFERENCES CITED

Allais, M., 1957, Method of appraising economic prospects of mining exploration over large territories—Algerian Sahara case study: Management Sci., v. 3, no. 4, p. 285-347.

Arps, J. J., and T. G. Roberts, 1958, Economics of drilling for Cretaceous oil on east flank of Denver-Julesburg basin: AAPG Bull., v. 42, no. 11, p. 2549-2566.

Barouch, E., and G. Kaufman, 1974, Sampling without replacement and proportional to random size: unpub. ms.

Cochran, W. G., 1939, The use of analysis of variance in enumeration by sampling: Jour. Am. Statistical Assoc., v. 34, p. 492-510.

Cox, D., 1962, Further results on tests of separate families of hypotheses: Jour. Royal Statistical Soc., ser. B, v. 24, no. 2, p. 406-423.

———1969, Some sampling in technology, *in* N. L. Johnson and H. Smith, eds., New developments in survey sampling: New York, John Wiley and Sons, p. 506-527.

Crabbe, P. J., 1969, The stochastic production function of oil and gas exploration in mature regions: Operations Research Branch, Natl. Energy Board, unpub. memo.

Drew, L. J., 1972, Spatial distribution of the probability of occurrence and the value of petroleum; Kansas, an example: Mathematical Geology, v. 4, no. 2, p. 155-171.

———1974, Estimation of petroleum exploration success and the effects of resource base exhaustion via a simulation model: U.S. Geol. Survey Bull. 1328, 25 p.

Ericson, W., 1969, Subjective Bayesian models in sampling finite populations (with discussion): Jour. Royal Statistical Soc., ser. B, v. 31, p. 195-233.

Feller, William, 1966, An introduction to probability theory and its applications, v. 2: New York, John Wiley and Sons, 627 p.

Fisher, R. A., 1956, Statistical methods and scientific inference: London, Oliver and Boyd.

Jackson, O. A. Y., 1969, Fitting a gamma or lognormal distribution to fibre diameter measurements on wool tops: Applied Statistics, v. 18, no. 1.

Kaufman, Gordon, 1963, Statistical decision and related techniques in oil and gas exploration: Englewood Cliffs, N. J., Prentice-Hall, 307 p.

———1965, Statistical analysis of the size distribution of oil and gas fields, *in* Symposium on petroleum economics and evaluation: AIME, p. 109-124.

———and P. G. Bradley, 1973, Two stochastic models useful in petroleum exploration, *in* Arctic geology: AAPG Mem. 19, p. 633-637.

Krige, D. C., 1951, A statistical approach to some basic mine valuation problems on the Witwatersrand: Jour. Chemical, Metall., and Mining Soc. South Africa, v. 52, p. 119-139.

Mandelbrot, B., 1960, The Pareto-Levy random functions and the multiplicative variation of income: Yorktown Heights, N. Y., IBM Research Center Rept.

Matheron, Georges, 1955, Application des méthodes statistiques à l'évaluation des gisements: Annales des Mines, December.

McCrossan, R. G., 1969, An analysis of size frequency distribution of oil and gas reserves of Western Canada: Canadian Jour. Earth Sci., v. 6, no. 2, p. 201-211.

Palit, C. D., and I Guttman, 1973, Bayesian estimation procedures for finite populations: Commun. in Statistics, v. 1, no. 2, p. 93-108.

Prohkorov, Yu., V., 1964, On the lognormal distribution in geochemical problems: Jour. Applied Probability and Its Applications.

Rodionov, D. A., 1964, Distribution functions of element and mineral content of igneous rocks: Moscow, Izd. "Nauka" (in Russian).

Ryan, J. T., 1973a, An analysis of the crude-oil discovery rate in Alberta: Bull. Canadian Petroleum Geology, v. 21, no. 2, p. 219-235.

———1973b, An estimate of the conventional crude-oil potential in Alberta: Bull. Canadian Petroleum Geology, v. 21, no. 2, p. 236-246.

Uhler, R., and P. G. Bradley, 1970, A stochastic model for determining the economic prospects of petroleum exploration over large regions: Jour. Am. Statistical Assoc., v. 65, p. 623-630.

Sharp, K., 1969, Lognormal vs. gamma distribution: Energy Resources Conservation Board, unpub. memo.

Zellner, A., 1960, An introduction to Bayesian econometrics: New York, John Wiley and Sons.

Assessing Regional Oil and Gas Potential [1]

D. A. WHITE,[2] R. W. GARRETT, JR.,[2] G. R. MARSH,[2] R. A. BAKER,[3] and H. M. GEHMAN[3]

ABSTRACT   Regional supply of potential oil and gas can be assessed by projecting growth of known fields, by extrapolating discovery rates in partially drilled sedimentary rocks, and by geologically analyzing undrilled sedimentary rocks. The *probable* supply from future growth of existing fields is estimated by projecting past year-to-year revision ratios that reflect the known pattern of reserve additions. The *possible* supply from new fields in partially drilled rocks is estimated by extrapolating historical discovery rates expressed as oil-equivalent barrels found per foot of new-field wildcat drilling. (This method applies only to maturely explored areas and depths where the discovery rate is generally declining.) The *speculative* supply from new fields in undrilled areas or depths is estimated by geologic analysis and comparison. As one example, after assessing the probable and possible gas, we estimate the deep speculative gas potential of south Louisiana by multiplying predicted volumes of sandstones, in cubic miles, by related yields, in cubic feet of gas per cubic mile of sand.

The results of each assessment are presented as a probability curve showing the existence chance for each amount (or more) of potential gas. The range of values on the curve reflects the combined uncertainties in the input factors of our estimate. Basic assumptions, interpretations, and quantities are specified so that validity may be checked by others. We account for geologic risk and any other anomalies affecting the outcome. Where feasible, each different play, area, or category is assessed separately using different approaches if necessary. Assessment curves are then added by Monte Carlo simulation. Our ultimate assessment goal is to make realistic judgments based on geologic fundamentals and experience.

INTRODUCTION

Three methods of assessing the future oil and gas potential of a region are (1) projecting the growth of existing fields, (2) extrapolating the oil-equivalent barrels discovered per wildcat foot in partly drilled sedimentary rocks, and (3) geologically analyzing undrilled sedimentary rocks. The first two methods apply only to areas with considerable drilling and discovery histories; if used with care, such methods can give the most realistic appraisal of mature exploration areas or depths. Only the undrilled or sparsely drilled section is assessed by geologic analysis and comparison.

Assessment results are presented as a probability curve. For example, Figure 1 shows on the vertical scale the respective existence chances of the different ·amounts (or more) of potential hydrocarbons (over and above proved reserves) shown on the horizontal scale. This particular curve is the final summation of the three assessment approaches outlined above. We have used the potential gas of inland south Louisiana for illustration, because it is important to test methods in actual geologic settings with real assessment challenges. Our answers are not the most complete or the ultimate ones for this area, however. They can be improved by use of more data and more time. Our emphasis is on the methods themselves, together with the underlying assumptions, interpretations, and quantities required.

The range of results on Figure 1 reflects the uncertainty in our estimates. We read from the curve that there is a 100-percent chance

[1] Manuscript received, January 8, 1975.
[2] Exxon Company, U.S.A., Houston, Texas 77001.
[3] Exxon Production Research Company, Houston, Texas 77001.

The authors thank these companies for permission to publish. We also thank G. C. Grender, J. W. Harbaugh, D. C. Skeels, Delma Smith, and R. I. Swanson for helpful suggestions.

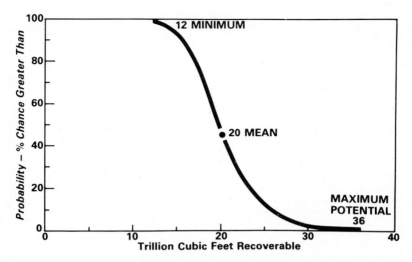

FIG. 1—Assessment of potential gas for inland south Louisiana, representing summation of *probable*, *possible*, and *speculative* estimates.

that at least a minimum of 12 Tcf of recoverable gas exists within the area, over and above the current proved reserves. There is a 45-percent chance that the mean average of 20 Tcf, or more, exists. There is an almost zero chance (1 in 500) for the existence of the maximum potential of 36 Tcf or more. (Potentials of 12, 20, and 36 Tcf equal about 340, 570, and 1,020 billion m^3, respectively.)

Such a curve conveys much more information than does a single number. A steep curve shows a very narrow range in the possibilities. A flat curve reflects a wide range of uncertainty about the outcome. It is the average, not the maximum, that should be added to the averages from other areas to get a realistic idea of a larger region's potential hydrocarbon supply. Direct addition of several maximum potentials gives an unrealistic and unattainable answer from a probability standpoint. Thus, a single-number assessment with no probability specified is not a good assessment. Also, because of all the unknowns and risks inherent in oil and gas exploration, the probabilistic way of expressing hazardous predictions has much practical appeal.

Most methods of building either local or regional assessment curves involve multiplying several factors that give answers in barrels of oil or cubic feet of gas. Ranges of values are assigned to each factor, depending on the uncertainty involved, and many answers are calculated by multiplying many different possible combinations of factors together in a Monte Carlo simulation. The many answers are then ordered by increasing size and plotted against the percent of the total number of answers that are larger than each given size.

Local (prospect) assessment methods may use (1) the basic reservoir-trap volume and its potential hydrocarbon fill, (2) "pay" factors (area-thickness-recovery), (3) the paleodrainage yield of hydrocarbons from source rocks, (4) special empirical relations (e.g., oil and gas versus sandstone/shale ratio), or (5) geochemical material balance of hydrocarbons generated, migrated, and trapped. Regional methods may use (1) summation of prospects or plays, (2) trap-size or field-size distributions, (3) Delphi average of expert opinions, (4) direct look-alike reserves comparisons, (5) areal hydrocarbon yields, (6) volumetric yields of total sediments, (7) fine(source)-facies yields, (8) coarse(reservoir)-facies yields, (9) projection of field growth rates, or (10) extrapolation of discovery rates. We illustrate only the last three because they are suited to our example area. The best methods are those that take into account most realistically the geology and experience in the area. Each approach must be carefully tailored to any special requirements and to the data available. Each result must be carefully weighed and adjusted, if necessary, for any geologic risks or other factors not previously taken into account.

DEFINITIONS

For the purposes of our assessment of south Louisiana, we use the following definitions of three categories of potential gas supply. These are modified from the definitions given by the Potential Gas Committee (1973) in order to relate to the three different methods used in making our estimates.

The *probable supply* results from growth of existing conventional fields. It is estimated from historical revision ratios. These ratios are determined by crediting back subsequent reserve additions to each field's discovery year. The method accounts for future extensions, revisions, and new-pool discoveries that are comparable to past experience. Any anticipated growth not comparable to past experience must be assessed separately.

The *possible supply* is from new fields in sedimentary rocks that have been partially drilled. It is estimated from the extrapolation of historical discovery rates. This extrapolation applies only to mature areas and depths where the discovery rate is declining. Preliminary studies (mainly in Western Canada) suggest that, with modern technology, most of the big fields are found by the time there is about one wildcat per township (36 sq mi or 93 km^2). After this, the discovery rate, expressed as equivalent barrels of oil found per foot of wildcat drilling, is apt to decline. The reserves to be discovered by future new-field wildcats within this partially drilled volume of sedimentary rocks are represented in our extrapolation; but the potential reserves in undrilled areas or depths are not, and these must be assessed separately.

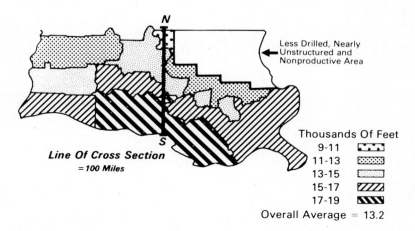

FIG. 2—Partially drilled sedimentary volume to various depths in south Louisiana, representing average total depths of deepest wildcat well in each township.

Thus the remaining *speculative supply*, from new fields in sedimentary rocks not drilled to a density sufficient to locate the big fields, has been estimated by geologic analysis and comparison.

SOUTH LOUISIANA SETTING

To separate geologically the speculative from the probable plus possible estimates, we assumed the "partially drilled" sedimentary volume to be that penetrated by at least one well per township. Accordingly, we sorted out the deepest wildcat well in each township and averaged all the total depths of these wells by parish. The map of south Louisiana (Fig. 2) shows that some of the northern parishes have been partially drilled to little more than 9,000 ft (2,740 m), whereas the average for the south-central coastal parishes is about 18,000 ft (5,480 m). The overall average depth, excluding the less-drilled, nearly unstructured, and nonproductive area in the northeast, is 13,200 ft (4,020 m).

A meaningful check on the depth currently separating the speculative from the probable plus possible is the average depth of deepest known production. The average producing depths of the more important new fields from 1968 to 1973 is 13,300 ft (4,050 m). (For an example of the data, see Stevens and Callahan, 1974, p. 1626.)

Figure 3 shows how these well and discovery depths relate to the geology and to the remaining speculative exploration opportunities. This north-south cross section cuts the center of the area. Although simplified and stylized for presentation, the thicknesses, rock types, and present production are carefully scaled from data given in papers by Mason (1971) and others. These papers are from a study by the National Petroleum Council, published as AAPG *Memoir* 15. A wealth of information about south Louisiana is also contained in AAPG *Memoir* 9 (Lafayette and New Orleans Geological Societies, 1968.)

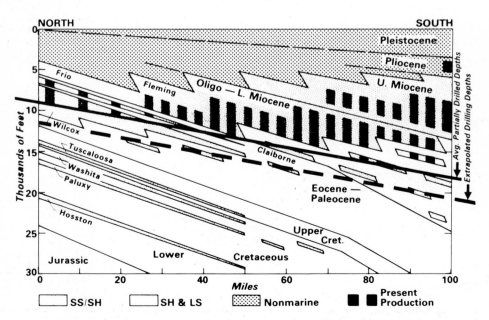

FIG. 3—North-south stratigraphic cross section, south Louisiana, prepared from data taken from Mason (1971) and others (see Fig. 2 for location).

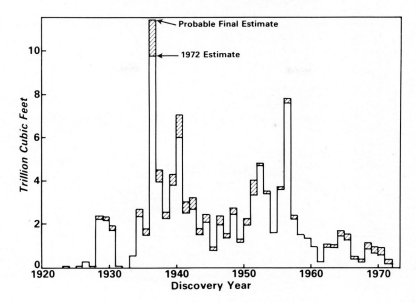

FIG. 4—Total gas discovered each year in south Louisiana, showing 1972 published estimate together with *probable* additional growth estimated in this study.

The sloping, solid black line on Figure 3 represents the current average depths of partially drilled sedimentary volume determined by parish. Known oil and gas occurrences lie mainly above this line. Our extrapolations to assess probable and possible hydrocarbon supplies take into account future pool and field discoveries by continued exploratory drilling in the sedimentary volume above the extrapolated dashed line. Our speculative assessment covers the volume below the dashed line.

ESTIMATING THE PROBABLE SUPPLY

The probable future supply from anticipated growth of known conventional fields is projected from the historic annual rate of change in reserve additions. The results for south Louisiana are shown on Figure 4 as the "probable final estimate," which is additional to the estimate of proved ultimate recovery (American Gas Association, 1973). The total proved ultimate recovery is about 93 Tcf of gas (75 Tcf nonassociated and 18 Tcf associated-dissolved) plus 9.5 billion bbl of oil and 3.0 billion bbl of natural gas liquids. We have made separate analyses for the newer fields discovered during and since 1953 and the older fields discovered before 1953.

The probable reserves added for newer fields (Fig. 4) reflect the usual time lag required to develop and calculate new discoveries fully. Figure 5 charts the reserves growth with time for nonassociated gas in south Louisiana; it is based on the American Gas Association's yearly tables of reserve estimates credited back to the year of discovery. The first year's ratio is the original discovery estimate divided into the first revised estimate made 1 year later. The second ratio is calculated by dividing the first revision into the second revision and then multiplying by the previous ratio, and so on. Reserve estimates grow rapidly in the first 10 years. After about 20 years of revisions, the newer fields have reached their ultimate sizes, averaging 2.3 times their

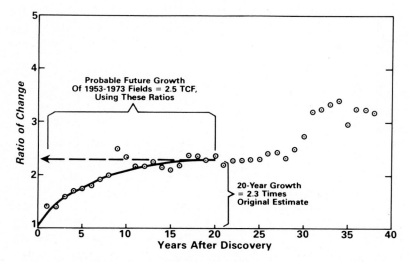

FIG. 5—*Probable* growth with time for nonassociated-gas reserves, south Louisiana, based on historical
record of generally increasing estimates relative to discovery-year estimate.

initial estimates. This relation, as shown by Marsh (1971), is used to
calculate the probable final reserves by multiplying each vintage of
calculated estimates by 2.3 and dividing by that year's revision ratio.
The additions of probable reserves of gas for newer fields in south
Louisiana are 2.5 Tcf. This includes 0.1 Tcf of associated gas calculated using the factors developed for oil.

The probable reserves added for older fields (Fig. 4) reflect current
growth mainly of the very large and complex fields found early in the
exploration of south Louisiana. Because this growth (the second hump in
Fig. 5) is not a predictable function of time since discovery, we have
analyzed it in a different way. The recent yearly changes in nonassociated-gas reserve estimates for old fields (Fig. 6) are given in the
American Gas Association's discovery tables. Although the data base is
small and the variations are large, we have made a least-squares fit of
the annual growth rate versus cumulative growth. This projection gives
a "most likely" additional 9.5 Tcf of gas within a judgmental range of
uncertainty from 5.5 to 15.5 Tcf. (There has been no growth of associated-
gas reserves, and none is projected.)

To complete our probable assessment, we add the 2.5 Tcf for newer
fields to the minimum, most likely, and maximum figures for older fields.
The "most likely" result of 12 Tcf agrees closely with the 11 Tcf growth
estimated by the Potential Gas Committee (1973) for existing fields above
15,000 ft (4,570 m) depth. The geologic complexities that permit such
growth are well illustrated in a paper by the Lafayette and New Orleans
Geological Societies (1968).

The final minimum, most likely, and maximum estimates are set in a
log-triangular distribution that gives the probability curve of Figure
7. The probable gas supply is thus assessed at about 8 Tcf minimum, 12
Tcf mean, and 18 Tcf maximum potential. Future increased incentives may
result in additional recoveries from marginal or "tight" gas reservoirs
not heretofore tapped in known fields. An adequate assessment of this
potential must be based on detailed geologic, engineering, and economic
studies beyond the scope of this report.

The chief uncertainties in these growth estimates are the usual difficulties in estimating reserves (Lovejoy and Homan, 1965), the problem of separating onshore from offshore reserves, the limited number of years having available successive reserve estimates by discovery year, and the sometimes arbitrary way of assigning discoveries to the proper year. In 1973, for example, large gas reserves that had been carried with younger fields for many years were abruptly taken out and credited back to older fields. This destroyed historical continuity and ruled out use of the 1973 data for most analyses. Because we account for growth of all fields, the particular assignment of reserves to any discovery year is far less critical than is the later consistent maintenance of the growth pattern.

Arrington (1960, 1966) pioneered the method of projecting growth rates of existing fields, as noted by Marsh (1971). Hubbert (1967) and Arps et al (1971) used similar approaches.

ESTIMATING THE POSSIBLE SUPPLY

Assessment of the possible supply from the partially drilled rocks is derived from an extrapolation of historical discovery rates to an economic limit. The generally declining rates for south Louisiana, expressed as probable final oil-equivalent barrels found per foot of new-field wildcat drilling, are shown for 1953-1973 in Figure 8. (Oil-equivalent barrels here include oil and total gas converted on an energy basis of 6,000 cu ft/bbl.) Also shown are the minimum, most likely, and maximum extrapolations.

The historical record of declining discovery rates on which our extrapolation is based includes no new frontiers; these are left for the speculative estimate. We assume that industry drilling will continue with a declining reward until a minimum economic limit of 6 bbl/ft is passed. Recent drilling costs in south Louisiana run about $30/ft, and other exploration costs add perhaps $30 more. At a new-oil (or gas-equivalent) price of $10/bbl, the point where exploration costs alone

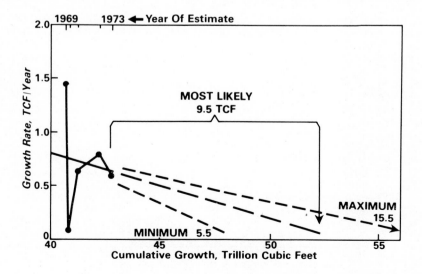

FIG. 6—*Probable* growth for nonassociated-gas reserves in pre-1953 fields, south Louisiana.

equal the undiscounted value of the reserves discovered is 6 bbl/ft. This is well beyond the point of profit for the industry effort. If future exploration costs rise in proportion to increasing prices, this economic limit would remain about the same. Furthermore, lowering this already low limit does not much alter the assessment.

The basic tenets of our analysis are that the number of barrels discovered per foot drilled varies randomly from year to year and that the future variation will be similar to the past. The analysis resembles a time-series Markov chain (Kendall and Stuart, 1968, p. 418), except that the random element is multiplied rather than added.

To smooth the high annual variability, this particular extrapolation uses 3-year averages of the discovery rate (Fig. 8). The starting point is 39 bbl/ft for 1971-1973. This number is multiplied in the computer by a factor randomly selected from an array of factors obtained by dividing each historical 3-year average by the preceding one. The first product is then itself multiplied by a similarly selected random factor. Multiplications are made for 100 successive values in the series, covering a span of 300 years. The whole process is repeated many times. The shortest of these chains—the one that first falls below 6 bbl/ft for 6 years running—is taken as the minimum, the longest chain is the maximum, and the median is taken as the most likely. Cumulative discoveries are estimated by multiplying the average discovery rate by the chain length in years by the historical average of 2.5 million ft (0.76 million m) of drilling per year. When the known discoveries for the years 1953 through 1973 are subtracted from the extrapolated ultimate discoveries at the economic limit, the resulting undiscovered minimum, most likely, and maximum values are 0.2, 0.9, and 3.6 billion oil-equivalent barrels, respectively (Fig. 8).

To convert to gas volumes, we multiply the above undiscovered oil-equivalent barrels first by 0.85, to allow for the recent-discovery proportion of 15 percent oil, and then by 6,000 cu ft/bbl. The results are set in a log-triangular distribution that gives the probability curve of Figure 9. The possible supply is thus assessed at about 1 Tcf minimum, 5 Tcf mean, and 16 Tcf maximum potential.

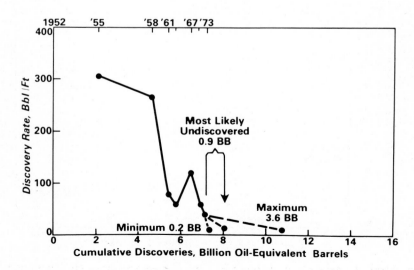

FIG. 7—Assessment of *probable* gas potential from growth of existing fields, south Louisiana.

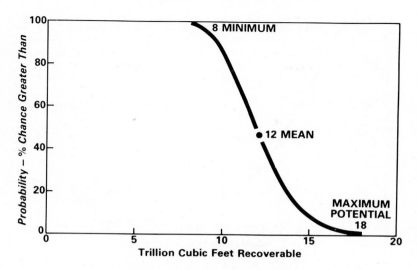

FIG. 8—Discovery-rate extrapolation for *possible* potential, south Louisiana.

Earlier workers laid the foundation for this type of extrapolation. Megill (1958) made a logarithmic projection of the relation between barrels discovered per successful wildcat and cumulative oil discoveries in Oklahoma and Kansas. Arps and Roberts (1958) used a linear extrapolation of the integrated form of the discovery-footage relation to estimate future discoveries in the Denver basin. Hubbert (1973), in several papers since 1956, has made similar extrapolations, the earlier ones being related to time. He later (1967) extrapolated the log of discovery rate versus cumulative footage, which is mathematically equivalent to the linear relation between discovery rate and cumulative discoveries shown on Figure 8. D. C. Skeels, in unpublished reports from 1962 to 1968 for Imperial Oil, Ltd., and Exxon Production Research Company, developed first a linear and then a logarithmic extrapolation to an economic limit of discovery rate (barrels found per foot of new-field wildcat drilling) versus cumulative discoveries.

Uncertainties in these projections include all those of the probable estimates plus further questions about the mathematical form of the extrapolation, the marked variations in annual discovery rates, the proper allocation of amount and type of drilling footage, the changing influence of economics, the fact that past history may not always be the key to the future, and the uncertain areas and depths involved.

Our method, developed by Baker, has advantages that appear to minimize, or at least clarify, these uncertainties. The extrapolation gives multiple answers and hence a probability distribution. Use of oil-equivalent barrels avoids arbitrary allocation of drilling to oil versus gas. The method is a direct model of past variations that requires no preconception of the linear-versus-logarithmic relation of discovery rate to cumulative discoveries. Indeed, our results give independent support to Hubbert's linear extrapolation, which implies (realistically, it appears) that future discoveries comparable to past ones are finite rather than unlimited.

We have specified an economic limit because every assessment of recoverable hydrocarbons has one, even though it is rarely identified. Any projection based on historic reserves, or any comparison using known

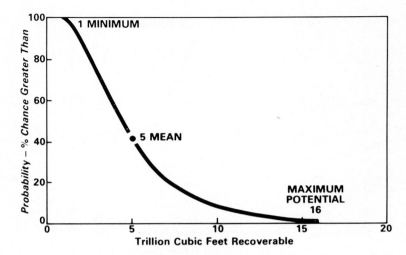

FIG. 9—Assessment of *possible* gas potential resulting from discovery-rate extrapolation in partially drilled sediments, south Louisiana.

hydrocarbon yields, is constrained by the economic climate in which those reserves were discovered and evaluated. Industry counts and records barrels or cubic feet only for economic successes, and it abandons the failures unmeasured. At the economic limit of 6 bbl/ft it would take about 50 million ft (15.2 million m) of new-field wildcat drilling to find the possible 0.9 billion oil-equivalent barrels. At 1 bbl/ft it would take another 50 million ft of such drilling to add only 0.2 billion bbl more. Nor would this be a scientifically pure number, because unusual or low-grade or very small deposits not included in the historical data cannot, of course, be adequately assessed by extrapolation. What we are trying to estimate is *attainable* potential.

Extrapolation of discovery rates also implies extrapolation of producing depths and other related quantities (D. C. Skeels, personal commun., 1974). A projection of the producing depths of important new fields (AAPG data—see Stevens and Callahan, 1974) suggests that the average partially drilled depth will reach about 16,000 ft (4,880 m) by the time the possible potential is finally discovered (Fig. 10). This average depth is represented by the dashed sloping line on the cross section (Fig. 3) and is our boundary between the possible and speculative assessment volumes. Although further research is needed, this analysis appears to remove much of the uncertainty about the sedimentary volumes represented by the extrapolation.

ESTIMATING THE SPECULATIVE SUPPLY

The factors in the speculative assessment are the estimated undrilled volumes of sandstones, in cubic miles, and the related gas yields, in cubic feet per cubic mile of sand. Multiplying these factors together gives the desired answer in trillion cubic feet. The sand volumes in south Louisiana are derived from the geologic maps of unit thicknesses and sand percentages given in the National Petroleum Council (NPC) study (Mason, 1971, and others). The gas yields are based on comparison with the established trends. The sedimentary volume to be assessed lies below the extrapolated plane of average depth to which the rocks are partially drilled (Fig. 3), down to a cutoff at 30,000 ft (9,140 m). The basic assessment approach used here was originally developed by Gehman.

Sandstone Volumes

The chief remaining sandstone plays are the downdip wedges of the Oligocene-lower Miocene Frio and Fleming, the Eocene Wilcox, the Upper Cretaceous Tuscaloosa, and the Lower Cretaceous Washita, Paluxy, and Hosston units (Fig. 3). For each play, geologists who made the NPC study stress that the occurrence of adequate reservoir sandstone is the most critical factor governing the chances for commercial hydrocarbons at these depths. Traps and source rocks for the gas are much more apt to be adequate. Accordingly, we have based our assessment mainly on the probable occurrences, volumes, and quality of reservoir rocks. The coarse-facies volume here is a much more sensitive indicator of hydrocarbon potential than the gross sedimentary volume, which contains considerable shale. Although this approach is suitable for this geologic setting, it cannot be used everywhere. Other or modified approaches must be used where traps or source rocks are the more critical geologic controls of oil and gas occurrence.

Table 1 summarizes the assumptions upon which we base our estimates of sand volumes. (Also included is the Lower Cretaceous limestone play, whose coarse-grained facies can be treated in the same way as the sandstone plays.) The total inland area is 24,000 sq mi (62,000 km²), and the areas of many plays are most likely to be about half this, or 12,000 sq mi (31,000 km²; Fig. 3). However, it is possible that the sands may not extend this far—or that the turbidite sands may extend farther. We allow for this uncertainty by giving a minimum-maximum range from one quarter to three quarters of the whole area, or from 6,000 to 18,000 sq mi (15,500 to 46,500 km²). One special play is the Eocene Wilcox, whose remaining speculative potential lies chiefly in the undrilled half of the northeast corner of the area, above the dashed line of Figure 3. Another is the Oligocene-Miocene, whose undrilled potential after explor-

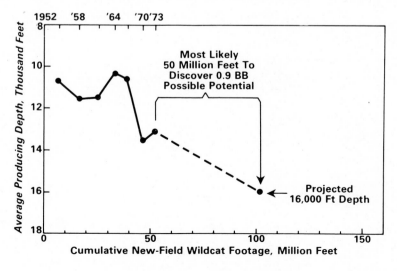

FIG. 10—Producing-depth extrapolation for important new fields, south Louisiana.

TABLE 1. SANDSTONE VOLUME FACTORS FOR SPECULATIVE ASSESSMENT OF SOUTH LOUISIANA
(MINIMUM - MOST LIKELY - MAXIMUM VALUES)

Unit	Area (1,000 Sq Mi)			Thickness (Avg. Ft)			Sand (%)			Volume (Cu Mi Sand)		
	Min.	M.L.	Max.	Min.	M.L.	Max.	Min.	M.L.	Max.	Min.	M.L.	Max.
Oligocene-Miocene	2	4	6	1,000	2,000	3,000	5	10	20	20	150	680
U. Cretaceous	6	12	18	100	125	150	5	10	20	5	30	100
Subtotal										25	180	780
Eocene	1	2	3	1,500	1,750	2,000	10	20	40	30	130	450
L. Cretaceous												
-Sandstone	6	12	18	1,200	1,400	1,600	5	10	20	70	320	1,090
-Limestone	6	12	18	400	1,000	1,600	5	10	20	20	230	1,090
Subtotal										120	680	2,630

ation for the possible hydrocarbons presumably will occur only in a few scattered deep pockets.

The thicknesses of the gross sandstone-shale units as mapped in the NPC study range from the postulated zero edges to the maximums on the fringes of the area. The average is halfway between, and Table 1 gives the ranges of values that express our uncertainties about these average thicknesses. The NPC study suggests that somewhere between 5 and 20 percent of each of these gross units is actually sandstone. We have taken these values as minimum and maximum, respectively, and have placed the "most likely" geometrically between, at 10 percent. Finally, assuming interdependence of all variables, we multiply these sand percentages by the respective minimum, most likely, and maximum thicknesses and areas to get ranges of estimates for the sandstone volumes in cubic miles (Table 1).

Hydrocarbon Yields

Estimation of an applicable range of hydrocarbon yields is always a difficult step. In Table 1 we have grouped the Oligocene-Miocene plus Upper Cretaceous sedimentary volumes separately from the Eocene plus Lower Cretaceous volumes. The reason, as suggested indirectly by Mason (1971), is that the historic yields of the latter group are only about one fourth those of the former group.

For the ultimate-yield estimate, the most important reference point is the yield of the currently productive rocks in south Louisiana, which is mainly the yield of the Oligocene-Miocene. This yield is calculated in Figure 11. We add the total cumulative production and proved reserves (93 Tcf of gas and 12.5 billion bbl of liquids) to the previously derived averages for probable (12 Tcf and 0.4 billion bbl) and possible (5 Tcf and 0.2 billion bbl) assessments, including all oil, gas, and natural gas liquids expressed as oil-equivalent barrels. We divide this sum of 31.5 billion bbl by the extrapolated partially drilled volume of sandstone, excluding the upper nonmarine unit shown on Figure 3. This volume is calculated by multiplying the total area of 24,000 sq mi by the average gross thickness of 9,000 ft, or 1.7 mi, and then multiplying by the average content of sandstone, 25 percent. The resulting yield is 3.1 million

• Ultimate Yield of Known Productive Sandstones

$$= \frac{\text{Production + Proved Reserves + Probable + Possible}}{\text{Cubic Miles of Sandstone}}$$

$$= \frac{31.5 \text{ billion oil-equivalent bbl}}{24,000 \text{ cu mi} \times 1.7 \text{ mi thick} \times 0.25 \text{ SS}} = 3.1 \text{ million bbl/cu mi SS}$$

$$= 16 \text{ Bcf} + 500,000 \text{ bbl NGL/cu mi SS}$$

• Speculative Yields for Deep Oligocene-Miocene

Discount Factor	Gas yields (Bcf/cu mi SS)		
	Minimum	Most Likely	Maximum
1/2	4	8	16

FIG. 11—Hydrocarbon yield factors for speculative assessment of south Louisiana.

oil-equivalent barrels per cubic mile of sandstone. If this yield were all gas and natural gas liquids, as we believe it will be at great depth, it would amount to about 16 billion cu ft of gas plus 500,000 bbl of natural gas liquids per cubic mile. (The conversion factor used throughout is 6,000 cu ft of gas per oil-equivalent barrel; natural gas liquids are assessed using the historic factor of about 30 bbl per million cubic feet of gas.)

In calculating this yield, we have tried to avoid three common pitfalls. First, estimated future potential is included, so that the ultimate comparative yardstick will not be too short. Second, the net sand volume for the yield is determined in exactly the same way as the sand volumes of rocks being assessed. Third, all hydrocarbons are accounted for, not just gas alone. (Consideration of the oil in place that theoretically could be converted to gas at great depth does not make much difference, because the conversion efficiency of 50 percent balances out the oil-recovery efficiency of 50 percent.)

How does this yield relate to the deep undrilled plays? The common prediction of the geologists who made the NPC study is that the deep accumulations cannot be as good as the known shallower ones. For the Oligocene-Miocene, Tipsword et al (1971, p. 838) wrote, "...the individual fields generally will be smaller....Most of the future discoveries should be associated with subtler, smaller, and deeper traps." For the Eocene, Lofton and Adams (1971, p. 860) stated: "Massive sand fracturing may be necessary to stimulate production." For the Upper Cretaceous, Holcomb (1971, p. 899) expressed the opinion that, "The unexplored part...in South Louisiana has poor potential with one possible exception. Where Woodbine [Tuscaloosa] sandstone units are turbidites..., prospects are fair." For the Lower Cretaceous, Rainwater (1971) wrote: "No major gas or oil accumulations have been discovered in Washita strata..." (p. 925); "...no porous beds are present in the proximal prodelta facies" (of the Paluxy sandstones—although Rainwater considered the deep limestones prospective; p. 922); and "The Hosston probably has the greatest potential..." (p. 907—but the Hosston is extremely deep over most of south Louisiana; Fig. 3).

Thus, much of our predicted deep sand volume, even if present, is apt to have greatly diminished porosity and permeability, with resulting lower recoveries. Furthermore, there will be high-pressure problems and other difficulties and expenses for finding and exploiting of the smaller flank traps to the south and the stratigraphic traps to the north.

For a logical tie point, we assume that the maximum yield for our speculative Oligocene-Miocene sand volume is the shallower known ultimate yield of 16 Bcf/cu mi. Now we must decide just how much worse than that our "most likely" yield will be. This is a difficult judgmental task, but we can at least specify how we make that judgment. Our opinion is that the finding and exploitation difficulties at great depth will probably eliminate small fields, and that recoveries from large fields will be cut by about half because of poor porosity and permeability. So we set the "most likely" at one half the maximum. The yield could be much less, so we set the minimum at one half the "most likely." We prefer to keep the three numbers geometrically related, because hydrocarbon yields have a truncated lognormal distribution, which simply means that there are many more little yields than big ones.

In selecting these yield factors, we have tried to be realistic in allowing for obvious geologic, technologic, and economic risks. Yet we have also recorded in our assessment what a reasonable maximum could be if there were no special problems or risks.

TABLE 2.   INPUT FACTORS FOR SPECULATIVE
ASSESSMENT OF SOUTH LOUISIANA

	Minimum	Most Likely	Maximum
For Oligocene-Miocene + Upper Cretaceous			
Sand volume (cu mi)	25	180	780
Gas yield (Bcf/cu mi)	4	8	16
For Eocene + Lower Cretaceous			
Sand volume (cu mi)	120	680	2,630
Gas yield (Bcf/cu mi)	1	2	4

Assessment

Table 2 summarizes our assessment input reflecting all prior analyses
and judgments. The sand-volume ranges are the same as described herein.
The Oligocene-Miocene yields reflect the geometric one-half discount
from the maximum ultimate known yield.   The Eocene - Lower Cretaceous
yields are set across the board at one fourth the Oligocene-Miocene yields,
in line with the conclusions of the NPC study.

For each geologic grouping, our assessment in trillion cubic feet
results from multiplying the specified volumes by the specified yields
(Table 2). Each range of values reflects the uncertainties in our estimates. We are also uncertain as to whether a low yield will go with a
high volume, or vice versa. To solve this problem, we have the computer
cover all reasonable possibilities, using a Monte Carlo simulation.

Our results (Fig. 12) are shown in the probability-curve format. For
the Oligocene-Miocene plus Upper Cretaceous, there is a 100-percent chance
of 0.2 Tcf or more, a 40-percent chance of 1.5 Tcf or more, and virtually
no chance of more than 8 Tcf.

Also on Figure 12 is the probability curve for the lower-yield sediments plus a curve combining the two assessments. In summing, we assume
that these are independent plays—that is, one may prove to be large and
the other small, or vice versa. Therefore, we use a Monte Carlo summation. Note that the means add directly, but the minimums and maximums
do not. The maximum sum is not as big as the direct addition of the two
individual maximums. This indicates that there is no reasonable chance
for such unlikely events as the occurrence together of the two maximum
potentials.

To summarize, we have said that, for all these deep rocks, we are
quite sure there will be at least 0.6 Tcf of gas recoverable, there will
be no more than 12 Tcf, and a realistic average estimate is 3 Tcf. We
have covered a large number of possibilities and have tied them to past
experience, but we still may not have covered them all. Our geologic
data for this assessment were not very detailed. New data or new ideas
may change our assessments of the minimum, most likely, or maximum
possibilities. Nevertheless, one of the great strengths of what we have
done is to lay out all the basic assumptions, recording our uncertainties

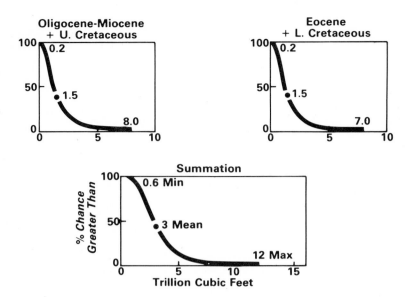

FIG. 12—Assessment of *speculative* gas potential in undrilled sediments, south Louisiana.

with specific ranges of values. Anyone can check out the logic and the reasonableness of the numbers to his satisfaction or dissatisfaction, and perhaps can suggest changes that will lead to better assessments.

SUMMARY

Our total assessment (Fig. 1) is the Monte Carlo summation of the probable (Fig. 7), possible (Fig. 9), and speculative (Fig. 12) assessments. If we have done our job well, the true answer will fall somewhere along the curve of Figure 1. If we were exploring many similar areas, we would expect the average of all their gas potentials to be near the average of this curve. Our aim has been to be pessimistic only for the minimum, to be optimistic for the maximum, and to be as realistic as possible on the average. We have striven to make effective use of geologic analysis of fundamentals as well as comparison with experience.

REFERENCES CITED

American Gas Association Committee on Natural Gas Reserves, 1973, Reserves of crude oil, natural gas liquids, and natural gas in the United States and Canada, and United States productive capacity as of December 31, 1972: Am. Gas Assoc., Am. Petroleum Inst., Canadian Petroleum Assoc., v. 27, p. 89-251.

Arps, J. J., M. Mortada, and A. E. Smith, 1971, Relationship between proved reserves and exploratory effort: Jour. Petroleum Technology, June, p. 671-675.

——and T. C. Roberts, 1958, Economics for Cretaceous oil on east flank of Denver-Julesburg basin: AAPG Bull., v. 42, p. 2549-2566.

Arrington, J. R., 1960, Predicting the size of discoveries—key to evaluating exploration programs: Oil and Gas Jour., v. 58, no. 9.

——1966, Estimation of future reserve revision in current fields, *in* Symposium on economics and the petroleum geologist: West Texas Geol. Soc. Trans., p. 16-30.

Holcomb, C. W., 1971, Hydrocarbon potential of Gulf Series of Western Gulf basin, *in* I. H. Cram, ed., Future petroleum provinces of the United States: AAPG Mem. 15, v. 2, p. 887-900.

Hubbert, M. K., 1967, Degree of advancement of petroleum exploration in United States: AAPG Bull., v. 51, p. 2207-2227.

——1973, Survey of world energy resources: Canadian Mining and Metall. Bull., July, p. 1-17.

Kendall, M. G., and A. Stuart, 1968, The advanced theory of statistics: New York, Hafner Publishing Co., v. 3, 552 p.

Lafayette and New Orleans Geological Societies, 1968, Geology of natural gas in south Louisiana, *in* B. W. Beebe and B. F. Curtis, eds., Natural gases of North America: AAPG Mem. 9, v. 1, p. 376-581.

Lofton, C. L., and W. M. Adams, 1971, Possible future petroleum provinces of Eocene and Paleocene, Western Gulf basin, *in* I. H. Cram, ed., Future petroleum provinces of the United States: AAPG Mem. 15, v. 2, p. 855-886.

Lovejoy, W. F., and P. T. Homan, 1965, Methods of estimating reserves of crude oil, natural gas, and natural gas liquids: Resources for the Future, Inc., Baltimore, Johns Hopkins Press, 163 p.

Marsh, G. R., 1971, How much oil are we really finding?: Oil and Gas Jour., April 5, p. 100-104.

Mason, B. B., 1971, Summary of possible future petroleum potential of region 6, Western Gulf basin, *in* I. H. Cram, ed., Future petroleum provinces of the United States: AAPG Mem. 15, v. 2, p. 805-812.

Megill, R. E., 1958, The cost of finding crude oil in Oklahoma and Kansas: Oil and Gas Jour., May 12, p. 189-198.

Potential Gas Committee, 1973, Potential supply of natural gas in United States (as of December 31, 1972): Golden, Colorado, Colorado School of Mines Foundation, Inc., 48 p.

Rainwater, E. H., 1971, Possible future petroleum potential of the Lower Cretaceous, Western Gulf Basin, *in* I. H. Cram, ed., Future petroleum provinces of the United States: AAPG Mem. 15, v. 2, p. 901-926.

Stevens, E. M., and R. L. Callahan, 1974, Developments in Louisiana Gulf Coast in 1973: AAPG Bull., v. 58, p. 1621-1629.

Tipsword, H. L., W. A. Fowler, Jr., and B. J. Sorrell, 1971, Possible future petroleum potential of lower Miocene - Oligocene, Western Gulf basin, *in* I. H. Cram, ed., Future petroleum provinces of the United States: AAPG Mem. 15, v. 2, p. 836-854.

False Precision in Petroleum Resource Estimates[1]

HOLLIS D. HEDBERG[2]

It is good that there are persons who are willing to come out with quantitative estimates for undiscovered petroleum resources. Such estimates make good after-dinner conversation (Hedberg, 1963) and, in giving a reasonable order of magnitude for such resources, they undoubtedly can be useful for economic and political purposes. However, such estimates can also be a real menace when, as too often happens, they give impressions of an accuracy beyond that which the meagerness of available knowledge warrants. You and I may understand that the last digits in an estimate of 2,272 billion bbl for the world's offshore recoverable petroleum resources are of no significance, and are only the results of carrying out far beyond any reasonable limits of error the arithmetic of the method of estimation or conjecture used. The public, however, may not understand this, and more than likely will correlate the degree of detail in the estimate figures with the degree of accuracy of the estimate. The nation then will rapidly proceed to build misguided plans and policies on what are highly dubious bases.

Until at least one well has been drilled and tested in a new area, no one can be entirely certain that the area will produce *any* petroleum in spite of the rosiest outlook; conversely, in spite of dim advance prospects, no one can be entirely certain that the area will not be a bonanza. So let me plead for more restraint in the resource estimates which are offered to the public. Preferably, let the estimates be made in round numbers—very round numbers—and if detailed figures must be used for the potential resources of new and undrilled areas, recognize the truth by making them in the form of ranges, with a zero always as the lower limit (e.g., 0-48 billion bbl for the recoverable resource off the Atlantic Coast), so that the public may get a true impression of the extent of assurance in the estimator's knowledge.

REFERENCE CITED

Hedberg, H. D., 1963, Paper *in* The role of national governments in exploration for mineral resources: Princeton Univ. Conference, Littoral Press, Ocean City, N. J., p. 4.

[1]Manuscript received, January 28, 1975.
[2]118 Library Place, Princeton, New Jersey 08540.

The Volume-of-Sediment Fallacy in Estimating Petroleum Resources [1]

HOLLIS D. HEDBERG[2]

The use of volume of sediment (sedimentary rocks) as an index for estimating the magnitude of petroleum resources can be justified in that petroleum generally originates in sediments, and without some volume of sediments the petroleum resources of an area are likely to be nil. Beyond this, however, there is little relation between the two, *unless* the promise of the sediments in question can be evaluated by qualifying them as to kind, organic content, lithologic character, lithogenetic character, thickness, age, etc, and unless other factors affecting the potential productivity of the area are taken into consideration. The use of unqualified volume-of-sediment figures, coupled only with figures for average productivity per unit volume, as a basis for estimating the resource volume of a region is all too commonly a delusion of the estimator and a snare for the uninformed public.

Many areas of the earth are covered with only a few hundred feet of sediments and their petroleum resource potential is zero; yet, because of the broad extent of these areas, the total volume of the sediments amounts to many thousands of cubic miles. Likewise, there are many areas with thousands of cubic miles of sediments that are too oxidized, too altered, too inaccessible, or too affected by other local conditions to be at all comparable with the sediments of oil-field areas. The bulk of the world's known oil comes from only a relatively few cubic miles of sediments, and the application of average productivity figures of any sort to the remaining volume is of dubious value. As I have said before (Hedberg, 1954, p. 1720), let me pick my cubic mile first and you can have all the rest!

Average wheat production in the world may be about 20 bushels per acre, but no reasonable person would attempt to calculate the wheat potential of the Canadian shield, or of the Sahara Desert, or of the Pacific Ocean, on the basis of area and some figure of world average productivity per acre. The use of such factors for estimating petroleum resources may be just as misleading.

REFERENCE CITED

Hedberg, H. D., 1954, World oil prospects—from a geological viewpoint: AAPG Bull., v. 38, p. 1714-1724.

[1] Manuscript received, January 28, 1975.
[2] 118 Library Place, Princeton, New Jersey 08540.

Undiscovered Petroleum Resources of Deep Ocean Floor [1]

K. O. EMERY [2]

ABSTRACT   Quantitative estimates of undiscovered resources of oil and gas beneath land areas have been prepared by many organizations and individuals on the basis of exploratory test wells and the presumed similarity of undrilled sedimentary basins to well-known ones. These estimates span a wide range of reliability. Undiscovered oil and gas resources beneath continental shelves probably are larger per unit area than those beneath the land, because the shelves are primarily marine depositional features, but the estimates of quantities beneath the shelves are even less reliable owing to fewer drillholes per unit area than on the land.

Least known of all potential oil and gas environments are the deep-water areas that include deep marginal basins, continental slopes, continental rises, and the deep ocean floor. Some analogs with the marginal basins are provided by similar basins that have been filled with sediments to form part of the continental shelf or even the adjacent land, but uplifted and accessible unmetamorphosed examples of continental slopes, continental rises, and abyssal plains are rare, if present at all. Marine geophysical data are available for these deep-ocean sedimentary environments, but drillhole samples do not exist other than those from the JOIDES Deep Sea Drilling Project, which intentionally avoided sites of oil potential to avoid possible pollution of the ocean. In addition, few of these drillholes penetrated as deep as 1,000 m into the bottom. Lack of suitable drillhole data means that quantitative estimates of undiscovered oil and gas resources of the deep ocean floor are meaningless—and, therefore, so are those for the whole world.

Offshore oil production began in the United States during 1896 when wells were drilled on piers projecting from shore at Summerland, California. This and most other offshore finds have been seaward extensions of oil or gas fields already known and producing on land. Even the diapiric fields in the northern Gulf of Mexico are just seaward extensions of trends and distributions on the adjacent land. With few exceptions, such as the younger oil in the North Sea and possibly that from Bass Strait off Australia, little oil and gas has come from fields that are unique or limited to the continental shelf. Annual crude oil production from the shelf off the United States diminished during 1971, 1972, and 1973 from 610, to 470, to 380 million bbl, respectively, owing to political-social-environmental restraints. Annual underwater production for the whole world during the same years generally increased from 3,200, to 3,100, to 3,500 million bbl, representing 18.2, 16.6, and 17.2 percent, respectively, of the total world production from both land and seafloor. As shown by Figure 1, production is widely distributed on the continental shelf. The chief areas avoided are those that are inhospitable owing to weather, politics, geology, or economics (listed in probable general order of decreasing limitation). Geologic limitations for the shelf are the same as those on land: absence of source beds (as on carbonate platforms), of reservoir beds (as on parts of deltas), of impervious caps, or of structural or stratigraphic traps.

Production and even exploration in deep-water ocean areas have lagged for the same reasons as for the continental shelf, but economic limitations have much greater relative control than on the shelf. One of the most interesting and highly potential geologic environments is that of small, deep marginal basins (Fig. 2). Such basins occur mainly in regions

[1] Manuscript received, December 23, 1974. Contribution No. 3425 of the Woods Hole Oceanographic Institution. Appreciation is due the Ocean Industry Program of W.H.O.I. for its support of this study.

[2] Woods Hole Oceanographic Institution, Woods Hole, Massachusetts 02543.

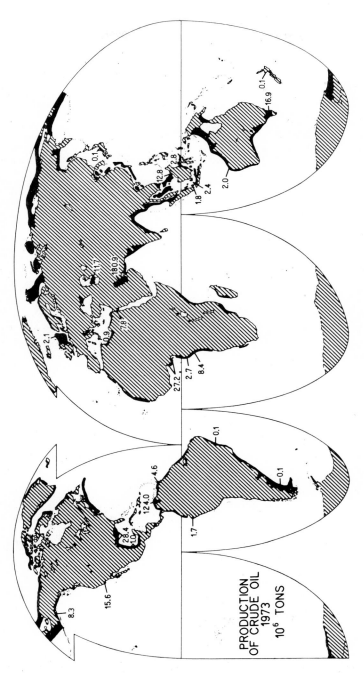

FIG. 1 — Areas in black represent most prospective parts of continental shelves; figures denote 1973 production of crude oil from fields on shelf.

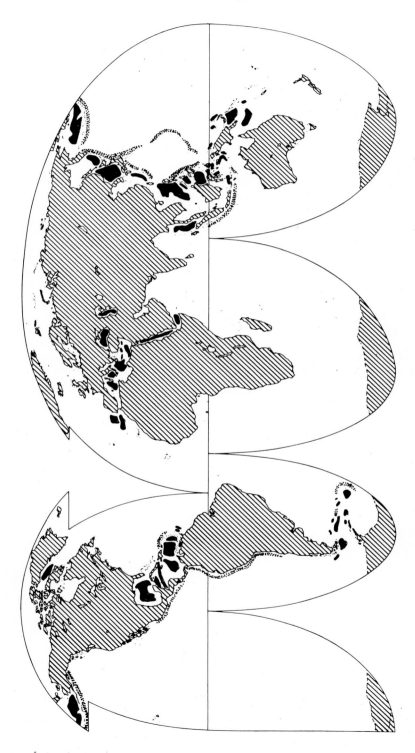

FIG. 2.–Distribution of small, deep marginal basins that have not yet become filled to overflowing with sediments.

of active tectonism and are confined largely to the junctions of conver-
gent crustal plates in the sense of seafloor spreading. The same high
relief that is represented by the small basins yields large supplies of
sediments that fill the basins, including both fine-grained organic-rich
sediments that can become source beds and coarse-grained sediments that
can become reservoir beds. Stratigraphic traps are abundant, and recur-
rent tectonism provides structural traps. Many of the basins eventually
become filled to overflowing with sediments, as illustrated by the Los
Angeles and Ventura basins in California. A limitation for oil potential in such basins is the probable restriction of thick sands (reservoir
beds) to the basins nearest shore that trap the sands from land and
prevent their reaching the next tiers of basins located farther seaward.
Owing to their inaccessibility to detrital sands, the far-offshore basins
of southern California, the East Indies, and the West Indies are unlikely
to contain large concentrations of recoverable oil.

Another deep-water environment having promise for petroleum is the
continental rise. As shown in Figure 3, the continental rise is restricted
essentially to continental margins of divergent crustal plates. Thus,
the rise would not contain marginal basins of the kind shown in Figure
2, but it could contain the generally smaller basins, now usually filled,
that were caused by tension during initial stages of divergent crustal
movement that broke protocontinents apart. In regions of convergence,
of course, the sediments that elsewhere could build a continental rise
are subducted in the long, narrow marginal trenches. Nearly half of the
earth's marine sediments occur in the continental rises, which border
most of the perimeter of the Atlantic, Indian, and Arctic Oceans. The
nature of these sediments at depth is conjectural because essentially
no drillhole data are available and because uplifted ancient representatives of continental-rise sediments are so altered by metamorphism that
they are nearly useless guides to the modern ones. The absence of test
holes in these areas by the JOIDES Deep Sea Drilling Project is due to
intentional avoidance for fear of possible pollution of the ocean by
escape of oil from uncontrolled drillholes. At present, we can suppose
only that most of the deep sediments of the continental rise contain very
little organic matter, owing to nearly complete oxidation in the oxygen-
rich deep-ocean water. We also suppose that sandstone beds generally
are absent. However, the upper continental rises probably receive abundant organic-rich silts and clays in the form of frequent landslides
from the relatively steep and unstable continental slopes that cross a
depth zone of water containing little oxygen and extending nearly ocean-
wide. Moreover, turbidity currents conducted through submarine canyons
probably deposit many potential reservoir sands on the upper continental
rises. Thus, the upper rises may well contain as-yet-unidentified resources of oil and gas.

Deep-water belts of diapirs are associated with some continental
rises and slopes in margins of divergent crustal-plate movement. Examples are common on both sides of the Atlantic Ocean (Fig. 4). The evaporite deposits that are the source of the diapirs formed during early
stages of continental separation when seaways were narrow and commonly
open at only one end, so that circulation of ocean water in the early
seas was restricted (as for the present Red Sea). Diapirs also occur
in some regions of nondivergent plate origin where water circulation was
impeded and evaporation was high with respect to precipitation and inflow
(such as the Mediterranean Sea during the late Miocene Epoch). In some
regions, such as western Africa south of Nigeria, the diapir belt occurs
not only under the continental rise but also underlies the continental
slope, shelf, and even part of the land area. The overlying sediments,
whose thickness and weight caused upward migration of the salt, usually

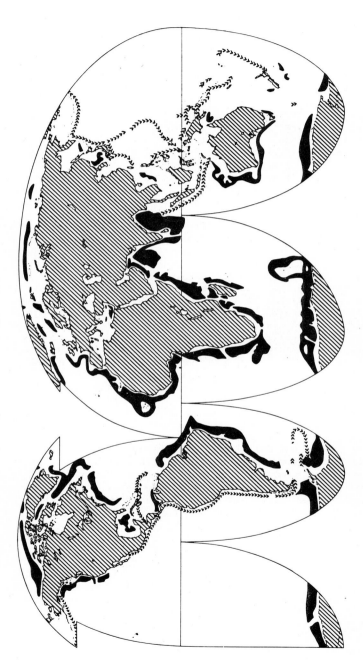

FIG. 3—Distribution of continental rises or large sediment prisms at base of continental slopes.

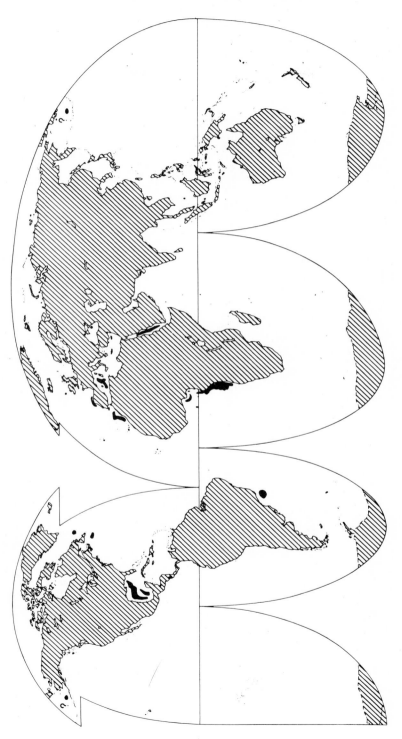

FIG. 4 – Distribution of deep ocean belts of diapirs.

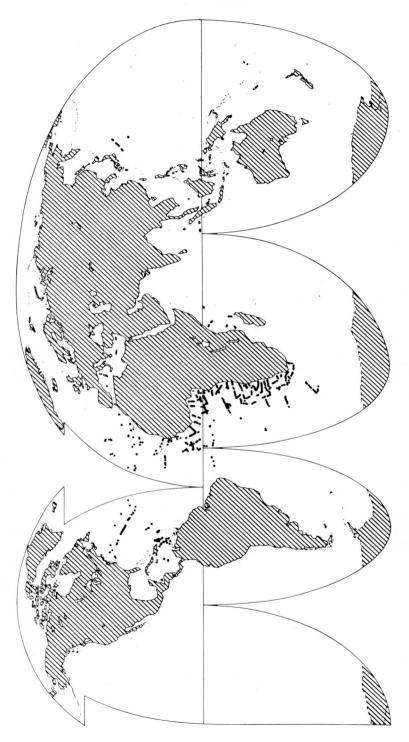

FIG. 5—Distribution of known "pagoda structures" (see Fig. 6) on deep ocean floor, as noted along about 1 million line-km of 3.5-kHz acoustic recordings, mostly in southeastern Atlantic Ocean.

contain sandy turbidites and enclose much organic matter brought by rivers; therefore, all of the requirements for large accumulations of oil and gas are likely to be present.

Another possible consideration is the presence of "pagoda structures" (Figs. 5, 6). These unusual and newly reported features are in the top 50-100 m of sediments in lower continental rises and abyssal plains. They are shown only in recordings of intermediate acoustic frequency in the vicinity of 3.5 kHz. Ordinary seismic reflection profiles made at 10-100 kHz miss them because these wavelengths can be longer than the thickness of the pagodas, and ordinary echo sounding at 12-16 kHz usually does not penetrate the sediments more than a few decimeters. At present, we know only that the pagodas are widespread in the Atlantic, scarce in the Pacific, and possibly present in the Indian Oceans—all at water depths of several kilometers and on otherwise smooth, flat ocean bottom. Speculation suggests that the cones between pagodas owe their origin to cementation of the sediments by some substance that reduces the differences in acoustic impedance of the sediment layers. The substances that appear most likely to fit the conditions of the environment and yet are not observable in core samples are gas hydrates, or clathrates, where the gas probably is methane. Whether or not these possible accumulations of natural gas are to be economic, the presence of the features and their unproved origin indicates that there are many things we do not know about oil and gas accumulations.

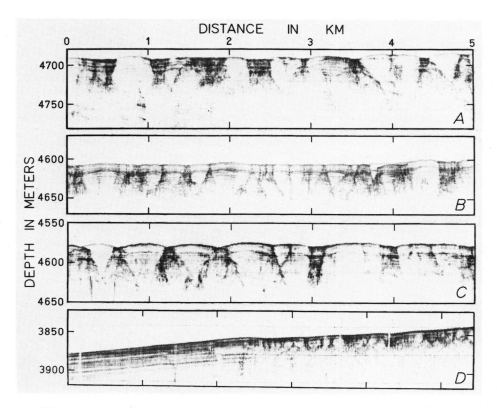

FIG. 6—Appearance of some 3.5-kHz acoustic recordings from off western Africa. Light-appearing cones may be marine sediments cemented by methane hydrates, or clathrates, whereas dark ones are typical of ordinary marine sediments of continental rises and abyssal plains.

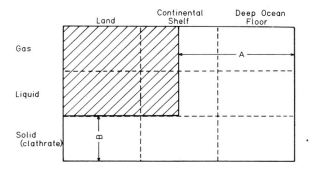

FIG. 7—Abstract diagram showing kind and distribution of petroleum hydrocarbon substances of earth. Hatched area denotes present area of major concern of oil companies.

As a general estimate, more than half the total volume of marine sediments of the earth lies beneath the surface of deep marginal basins, continental rises, and abyssal plains. Much, if not most, of the evaporites also occur beneath continental margins. Oil and gas have been found in a few JOIDES drill cores on the deep ocean floor, but, since not one hole intended for petroleum exploration has been drilled in the ocean floor beyond the continental slope, how can we possibly pretend to make an inventory of the oil and gas of the whole world? Only by making one or both of two assumptions can we believe that we are even close to making an inventory of the oil and gas potential of the world. One assumption is that there are no large volume accumulations of oil and gas on the deep ocean floor; is this assumption valid in the absence of a single test hole directed to that objective? The second assumption is that drilling and production from the deep ocean floor are inherently wasteful of energy—that is, the energy cost of recovery is greater than the energy to be recovered from the deep ocean floor. Will this assumption be valid a decade or two hence? If neither assumption is valid, in view of the evident future petroleum shortage we should be preparing to make a test investigation, but I know of none now being initiated in the United States. The major concern of oil companies is the oil and gas underlying the land and part of the continental shelves. Their estimates of undiscovered but recoverable resources of oil and gas can be valid for the whole earth only if (1) the deep ocean floor contains no recoverable hydrocarbons and (2) the gas hydrates postulated to be present there are found to be absent or uneconomic (A and B on Fig. 7). Do we know enough about A and B to believe that we can make this inventory?

Estimating Exploration Potential [1]

W. W. HAMBLETON, J. C. DAVIS, and J. H. DOVETON[2]

ABSTRACT   The Kansas Geological Survey has actively pursued a long-term interest in petroleum reserve estimation; the earliest Kansas reserve appraisals were published in 1896. Periodically updated reserve assessments have been issued routinely and are currently monitored by a computer file system. The major decline in Kansas crude oil production since the mid-1950s has placed a key emphasis on estimation of potential undiscovered reserves as a decision guide to future petroleum exploration in the state. Initial work in this area was undertaken by Griffiths, who applied the unit-value concept to the regional evaluation of potential mineral wealth across Kansas.

Techniques of regional resource evaluation may be related to two basic analytical models. In a causal predictive model, estimation of unknown reserves is derived as the effective outcome of causative basin parameters such as geometry and magnitude of source beds and potential reservoir formations. In contrast, an empirical predictive model entails the extrapolation of regional production history into the future, utilizing statistical time-series methods. The Survey is currently developing the Kansas Oil Exploration (KOX) Decision System incorporating features of both these models in a capability for estimation of unknown reserves in either local or regional areal contexts. Information sources of drilling histories and diagnostic geologic variables are analyzed interactively in a conditional probability framework geared to likelihood of undiscovered production.

The varying degree of oil exploitation across the state permits the application of statistical "experience" from more mature areas to the analysis of less-developed regions. Because the perception of basin characteristics undergoes a progressive evolution with the exploration history of a basin, the consequent dynamic character of key variables is selectively analyzed in a structured "re-experience" model of basin development. The production outcomes of wildcat areas are expressed as conditional probabilities drawn from contingency tables relating known production to perceived subsurface geology. These probabilities, which may be mapped areally, provide basic data for exploration decisions or economic analysis.

INTRODUCTION

Since the end of the last century, the Kansas Geological Survey has collated data on oil exploitation within the state and issued periodic reports on industry exploration and production statistics, together with projections of known reserves. The earliest Kansas reserve appraisals were published in 1896. More recently, the large data base associated with this activity has been monitored and synthesized by computer file manipulation. The time span of Survey reporting has covered much of the lifetime of Kansas as an oil province, which is now at a mature stage of development. State production has passed its primary peak and known crude oil reserves have declined since 1958 (Fig. 1).

At such a time, assessment of the location and magnitude of undiscovered reserves is a matter of special concern, both directly to local industry and indirectly to the state. The estimation of unknown reserves is inevitably a speculative process; if this were not the case, then oil exploration would not be universally recognized as the high-risk business that it is. However, such estimates are imperative, since they define the economic-reward infrastructure of the region under exploration. In turn, the monetary expectation dictates the degree and type of exploration activity commensurate with efficient discovery and exploitation.

Pioneer research work in Kansas resource evaluation was undertaken by Griffiths (1969), who utilized the unit-value concept in a regional

[1] Manuscript received, January 9, 1975.
[2] Kansas Geological Survey, The University of Kansas, Lawrence, Kansas 66044.

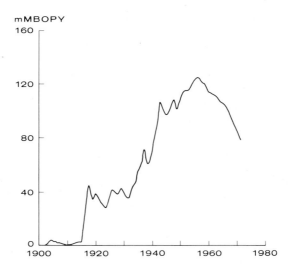

FIG. 1—Annual crude oil production in Kansas, in millions of barrels per year.

assessment of mineral wealth across Kansas. In the last 2 years, the
Survey has focused its attention on models specifically geared to prediction of undiscovered crude oil reserves in relatively immature areas
under current exploration by local industry. The uncertainties associated
with all phases of this problem strongly suggest that the best tools for
the job are techniques developed in statistics and economic decision
theory. Inferential statistics is closely linked with decision making
under conditions of uncertainty, which certainly apply to oil exploration.
In addition, the development of an exploration area is subject to both
the geologic factors that prescribe oil entrapment and the prevailing
economic environment. In view of these considerations, the Survey has
initiated work on the design of a statistical decision system that incorporates both geologic and economic factors. The system is modular in
structure, allowing both flexibility in operation and adaptability to
differing regional contexts.

The key prefatory step to the construction of a decision system is
the selection of an appropriate theoretical model that realistically
relates undiscovered reserves to observed parameters. Two basic choices
can be identified with either causal or empirical methods. In a causal
model, basin parameters of source-bed and reservoir-formation geometries
are analyzed concurrently with reconstructions of geologic history
relating to migration patterns and maturation processes. Such models
may be appropriate for virgin areas where the major information source
concerns the regional geologic framework of the basin and only limited
subsurface control is available. The approach is a conceptual ideal
that is difficult to apply convincingly, because basin parameters cannot
be estimated with any real degree of accuracy. A more fundamental objec-
tion, however, is the fact that concepts of petroleum generation and
migration are still a matter of scientific debate, so unequivocal basin
evaluations are the exception rather than the rule.

The alternative model of empiricism is most simply applied as the
extrapolation of production, where drilling histories are used as source
data for time-series estimation procedures that are evaluated on either
regional or local scales. A major drawback of this approach is that the

input data strongly reflect gross changes in the economic environment, such as wars, price legislation, and major shifts in market forces. Results of a model of this type are also restricted to estimations of magnitude rather than location of potential reserves; thus they function more as an indicator of the status of an oil province rather than a constructive guide for future exploration.

The Kansas Oil Exploration (KOX) Decision System employs a hybrid model implementing both causal and empirical elements in a design tailored for Kansas evaluations, but also applicable to many other areas. Although Kansas oil production is at a mature stage, the degree of exploitation varies across the state; therefore, local areas of advanced maturity are contrasted with neighboring regions in early stages of exploration. This characteristic enables statistical "experience" to be gained from exhaustive historical analysis of more mature areas; this "experience" can then be applied in a predictive model to adjacent immature areas. The causal content of the model is reflected in the choice of geologic variables and analytical methods together with the matching of "target" with "training" areas whose regional geology is judged to be effectively similar. The empirical phase of the methodology is contained in the historical "re-experience" analysis of the training area, where the dynamic evolution of perceived subsurface geology with field discovery is related to drilling-pattern development in time and space. Empiricism extends to the basic philosophy behind the system, which is heuristic, aiming to find the most efficient means of relating observed variables to unknown production in terms of conditional probability.

The following illustrative example highlights some of the main features of this approach in a simplified treatment commensurate with the limitations in scope of this paper. The example is "real" in the sense that the target area is under active exploration, a fact that should add interest to the outline of ideas presented.

APPLICATION OF KOX SYSTEM TO AN EXPLORATION STUDY

The Upper Pennsylvanian Lansing - Kansas City Group consists of interbedded limestone and shale units and is the primary oil-exploration target in northwest Kansas. The pattern of exploration in this area shows a diffusion from concentrated drilling activity in the southeast to relatively sparse well densities in the more northwestern counties. At a regional scale, oil fields are clustered on the main trend of the Central Kansas uplift and its northwesterly structural continuation as the Cambridge arch. A county area of current exploration interest and submature development was designated the target for analytical treatment. Differences in exploration maturity across the region enable the selection of a nearby relatively mature area on the southeast to be used as a training model. The training area was carefully chosen to have similar size, equivalent regional geology, and comparable local subsurface variability to the target area, as judged from available well control. The area is centered on Graham County (Fig. 2).

The structural block diagram (Fig. 3) is a three-dimensional representation of the top of the Lansing Group in the training area, as interpolated from the 2,758 wells that have been drilled. The surface is a broad central basin whose northern margin roughly coincides with the flank of the Central Kansas uplift. The basin is bounded on the east and west by positive structural blocks and is probably the product of late downwarping movements along persistent lines of weakness associated with the uplift. The distribution pattern of existing oil fields shows

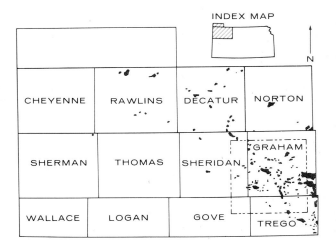

FIG. 2—Index map of northwestern Kansas region (shaded) which contains target area and Graham County training area.

a strong concordance with local structural highs (see Fig. 6) and suggests that Lansing structure may be highly diagnostic in the prediction of field locations.

Historical Analysis of Structural-Surface Perception

Relatively dense well control in the training area ensures that the interpolated structural surface is a close approach to the "real" Lansing surface. However, this was not the case in the past when considerably less well information was available. The perception of any geologic surface, both at regional and local scales, undergoes a dynamic evolution with time as the network of well control expands and diffuses across the area. Three time slices of a historical continuum are shown in Figures 4A, 5A, and 6A, which show the Lansing surface as it would have been perceived by interpolation from contemporary well control in 1937, 1952, and 1974. It should be emphasized that for each of these years the perceived structural surface would have been the major information source for exploration decisions executed in the following year.

The approximate realization of the "true" surface in 1974 allows hindsight analysis to be applied to earlier perceptions. During exploration development, the agreement between the perceived and true surfaces obviously varied considerably across the area. The perception of the surface at well sites is exact (assuming accurately picked tops) but becomes uncertain in intervening areas largely as a function of distance from control wells and the scale of subsurface variability. A quantitative assessment of the bounds of error associated with an interpolated estimate may be derived by hindsight analysis of the structural surface as perceived in 1952. Estimations of this surface at the locations of *post*-1952 wells (Z) at distances d from the nearest *pre*-1952 well were contrasted with their true surface values ($\overset{*}{Z}$). These errors may be expressed as standard deviations by the formula:

$$s_d = \sqrt{\Sigma\ (\overset{*}{Z} - Z)^2 / n}\ .$$

A plot of these error standard deviations versus distance from the nearest well (Fig. 7) shows a pronounced and systematic trend from absolute "certainty" about the Lansing elevation at a completed drilling site to a limiting constraint of "complete uncertainty." The sill of maximum indeterminancy is reached at a range of approximately 4 mi (6.4 km) from nearest control and is an isotropic measure of the average scale of local structural variability. The limiting uncertainty value is equal to the standard deviation of the entire surface about its mean value. At locations outside the determinancy range of any drillsite, contributory information from the set of control wells is too weak to assess the conditional variability of the fluctuations in the structural surface. In these areas, the error is effectively unknown and the standard deviation of the total surface provides the best estimate of the envelope of possible error. However, standard errors that are conditional on distance from control are applicable for locations within the 4-mi (6.4 km) range, as expressed by the functional relationship. (The "uncertainty" function shown in Figure 7 is a simple spatial autocorrelation model and has much in common with the more rigorously developed "Kriging" method popularized by Matheron [cf. 1963].)

The uncertainty function may be used to map the possible error in the interpolated estimates of structural configuration derived from the well-control array in the training area. Uncertainty contours are shown in supplementary maps with the time-slice series of Lansing surface perceptions in Figures 4B, 5B, and 6B. Superimposition of the two sets of contours permits an interpolated surface value to be linked with its computed error range at any location. (It may be assumed that the range of possible surface values at any point is described by a normal error curve whose first two moments are the estimate and the standard deviation.) The uncertainty associated with perception of the Lansing surface can be seen to diminish through the historical sequence from an early high level of uncertainty to a present-day pattern of relict high-uncertainty zones in limited areas of modern sparse well control.

Approximately 349 wells have been drilled in the target area to the northwest, and it is therefore at a similar stage of exploration development and perception as the training area was in 1952, when 352 wells had been drilled there. The two areas can be considered to be in parallel time streams of exploration activity. A map of the Lansing surface in the target area made from present-day well control (Fig. 8) shows a diffuse north-south ridge across the central part of the area, connecting positive structural features in the northeast and southwest. If local

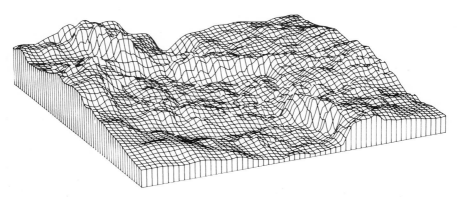

FIG. 3—Structural block diagram of top of Lansing Group (Pennsylvanian) in Graham County training area, based on data from 2,758 wells.

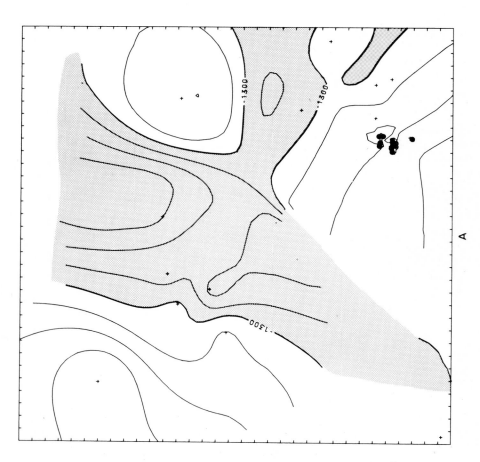

FIG. 4—A. Structure map of top of Lansing Group as perceived in 1937, when 31 wells had been drilled in Graham County training area. Producing wells are indicated by black dots, dry wells by crosses. C.I. = 20 ft. Areas below −1,300 ft are shaded. B. Estimated error associated with perceived structural surface. C.I. = ±20 ft possible error around structural-surface estimate. Tick marks along margins of both maps are at 1-mi intervals.

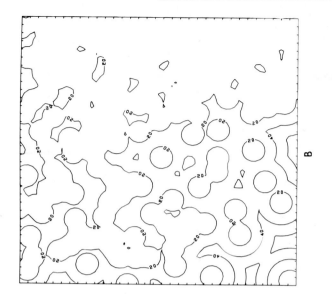
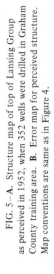
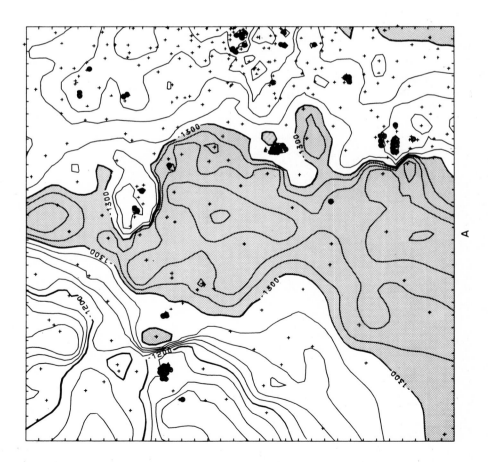

FIG. 5 – A. Structure map of top of Lansing Group as perceived in 1952, when 352 wells were drilled in Graham County training area. **B.** Error map for perceived structure. Map conventions are same as in Figure 4.

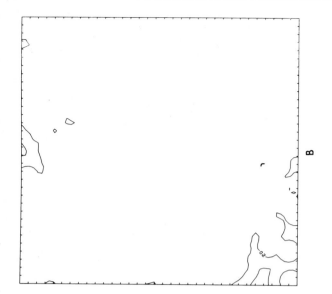

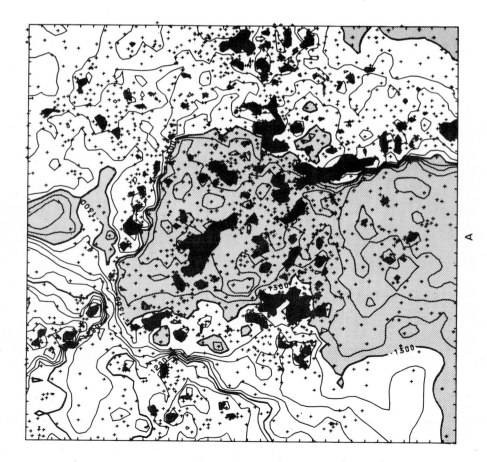

FIG. 6—A. Current (1974) perception of structure of top of Lansing Group in Graham County training area. Area now contains 2,758 wells. B. Error map for perceived structure. Map conventions are same as in Figure 4.

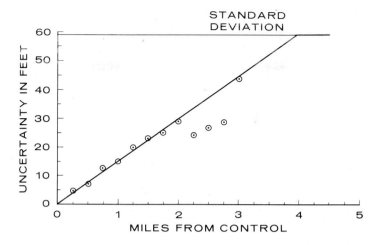

FIG. 7—Standard deviation ("uncertainty") of error in structural surface estimated from 1952 well control in Graham County training area, expressed as function of distance from nearest well control. Error increases with distance to a limit which is standard deviation of entire surface.

surface variability in the Lansing is effectively equivalent in training and target areas, then hindsight analysis of the training area provides a viable tool for predictions in the target area. Application of the uncertainty function to the target-area control network results in an error map that is a corollary to the interpolated surface-estimation map in Figure 8. The two maps are a necessary preliminary step in the analysis of probabilistic relations between Lansing structure and field locations.

Relations Between Structure and Oil-Field Location

Explorationists do not characterize structure in absolute terms of feet above or below sea level, but in relative terms such as anticline or syncline. Unfortunately, it is difficult to quantify these expressions of local shape, even though they are the elements of structure which control the entrapment of petroleum. Attemps have been made to define local structure qualitatively within cells (Prelat, 1974; Grender et al, 1974), but these are subjective assessments and hence liable to bias. Demirmen (1973a, b) devised a series of measures which are quantitative expressions of the rate of curvature of a surface around the point being classified. Other methods of measuring local structure include map filtering (Robinson et al, 1969; Robinson and Ellis, 1971), which passes only structures of specified dimensions, and separation of structure into "local" and "regional" components by trend-surface analysis.

Because trend-surface analysis is computationally the most straightforward of these methods and is already familiar and accepted by explorationists operating in northwestern Kansas, it was used to provide the measure of local structure. A polynomial trend surface tends to conform to the regional structural grain, leaving smaller features as residuals or deviations. At any point, either a well location or prospect site, local structure may be measured as the magnitude of the deviation from a trend.

A conditional relation between trend-surface residuals and oil occurrence can be found in the training area by calculating the trend as it would have been perceived at some prior time. It may be hypothesized that oil is preferentially associated with positive residuals; a contingency table can be constructed showing the ratio of success against

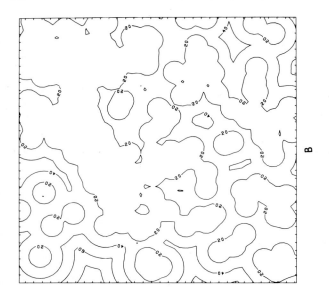

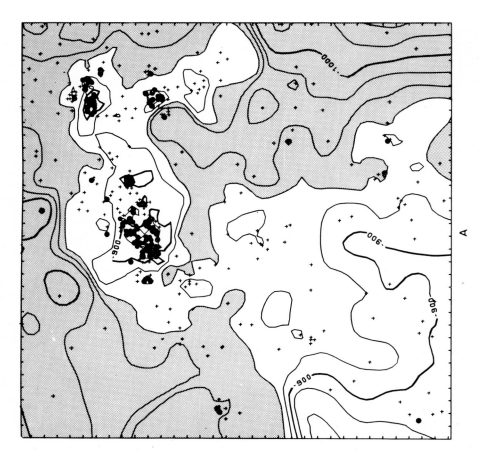

FIG. 8–A. Structure map of top of Lansing Group in target area as it is currently (1974) perceived. Area contains 358 wells. B. Error map for perceived structure. Map conventions are same as in Figure 4.

FIG. 9—Fourth-order trend-surface residuals calculated from well control available in Graham County training area in 1952. Positive residuals are shaded. C.I. = 20 ft.

residual magnitude for all wells subsequently drilled in the training area. Figure 9 shows fourth-order trend-surface residuals for the Graham County training area, calculated using all well information available through 1952.

Table 1 is a contingency table based on 2,333 wells drilled in the test area after 1952. This gives relative frequency estimates of the probability of success in drilling ventures on perceived structural residuals. The average proportion of successful wells in this set is 0.32; Figure 10, obtained from the contingency table, shows that the ratio is significantly higher in areas of high positive residuals and lower in areas of negative residuals.

The conditional probability distribution in Figure 10 suggests that trend-surface residuals could have been used directly to guide petroleum exploration in Graham County. However, the perception of residuals is subject to the same uncertainty as perception of the structure itself, because the residual map is a linear transformation of the structural surface. Therefore, the probability of success at any prospect site must be adjusted to take into account the uncertainty in residual magnitude, reflecting the relative density of well control in the vicinity, in exactly the same manner as the structural map itself was adjusted.

TABLE 1.   CONTINGENCY TABLE OF RESIDUAL MAGNITUDES, BASED ON
WELLS DRILLED IN GRAHAM COUNTY, KANSAS, AFTER 1952

Magnitude (ft)	No. Producing Wells	No. Dry Holes	Probability
>50	82	79	0.51
40 to 50	43	38	0.53
30 to 40	55	74	0.43
20 to 30	70	100	0.41
10 to 20	135	165	0.45
0 to 10	115	275	0.29
-10 to 0	114	263	0.30
-20 to -10	63	216	0.23
-30 to -20	24	135	0.15
-40 to -30	23	105	0.18
-50 to -40	11	75	0.13
< -50	7	66	0.10
Totals	742	1,591	0.32 (ambient probability)

At any point in the training area, the perceived trend residual may be
regarded as a point estimate of the true residual, and the uncertainty
as the standard error of this estimate. If the errors are assumed to
have approximately normal distributions, an adjusted probability of success
can be calculated. This is done by multiplying the probability of success
in a class of residual magnitudes by the probability that the residual
actually falls into that class. Summing these products across all classes
yields the adjusted probability of success for that location.

Figure 11 is a map of the probability of success for drilling ventures
in Graham County, as it would have been perceived at the end of 1952.
This illustration should be compared with Figure 5, which shows oil fields
known to exist at the time for which the probability map was calculated.
The probability map successfully predicts fields which were subsequently
discovered in the northwestern corner and southwest of the center of the
county. These discoveries are especially encouraging because no production had been found in either area as of 1952. The map also predicts
the multimillion-barrel Hoof field southeast of the center of the county,
although nearby production was already leading explorationists toward
its discovery. The probability map fails to predict the discovery of
the complex of fields in the structurally low region in the center of
the county. This reflects the low drilling density available in this
area in 1952, which resulted in relatively high uncertainty about the
local structural configuration.

Because the probability map is based on post-1952 well outcomes, it
might be expected that each category would contain exactly the number of
outcomes as in the original contingency table. However, this is not the
case because of the uncertainty of perception. Probabilities in rank
wildcat areas will be adjusted up or down toward the regional or ambient
probability. This is the level of indifference, or the success ratio
that an operator might anticipate if he were to drill at random, making
no attempt to select promising sites or to avoid unfavorable locations.
It also should be emphasized that the probability map will change as

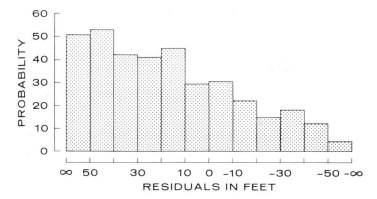

FIG. 10—Probability of discovery for drilling ventures in perceived trend-surface residuals of specified magnitude in Graham County. Probabilities based on data from 2,333 wells drilled in training area after 1952.

drilling continues in an area, reflecting not only an increase in perception, but also a decline in the ambient probability as the potential targets are successively tested. However, updated probability maps will continue to indicate the relative merits of potential sites within the area, even though these might become quite discouraging.

It also should be emphasized that this particular example is based on a two-state contingency table, so the probability map expresses only the likelihood of discovering oil. It is possible, given an adequate historical base, to construct a multi-state contingency table that will yield not only the likelihood of discovery, but also the likelihood that discoveries will be of specified magnitude. A series of probability maps can then be drawn, one for each category of field size. The sum of the probabilities on these maps will be equal to those shown on the probability map shown in Figure 11. Such an approach provides a way of estimating not only the desirability of a prospective area in terms of anticipated discoveries, but also allows quantitative estimates to be made of the volume of oil that will be found.

However, the objective of the KOX project is not to predict a hypothetical history for Graham County or to play the popular game of "what would have happened if. . . ." Rather, the purpose is to gain experience in a probabilistic sense, which can be applied to the target area as a guide for current and future exploration activity. Information from the Graham County training area has been combined with the currently perceived impression of geology in the target area to yield a map which can guide future exploration. Drilling is being conducted in the target area at the present time, and the outcomes of these wells are being predicted prior to their completion. This is the critical test of our research, and, if it is successful, it should significantly affect the pattern of exploration in Kansas. Obviously, no exploratory method will increase the amount of oil in the ground, but it may make the discovery of this oil more efficient and produce a resulting salubrious effect on the Kansas exploration industry.

CONCLUSIONS

The Kansas Oil Exploration Decision System was designed to help Kansas explorationists operate in the most efficient way possible. It exploits the varied rate of development in the state, taking probabilistic

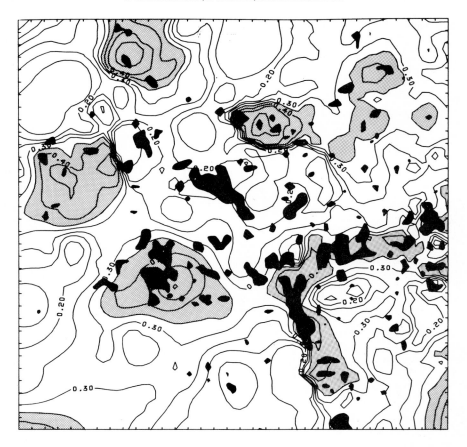

FIG. 11—Probability map for Graham County training area, based on structure as perceived in 1952. Contour interval is 0.05 probability of discovery of oil. Areas having probabilities higher than ambient level of 0.32 are shaded. All fields in training area are indicated in black. Compare with Figure 5 to determine association of post-1952 fields with areas of high probability.

"experience" from one area and applying it in another, just as geologists transfer their training and experience into new areas of operations. It also uses new and sophisticated methods of spatial analysis to assess the uncertainty which always accompanies any exploratory activity. Although KOX has been applied to one particular geologic province, using a single, specifically selected geologic variable, the methodology and operational philosophy are quite general. Any spatially continuous geological, geophysical, or geochemical variable or group of variables could be employed in any target area for which there is an adequate training area. We have tailored our application to fit the needs and resources of independents in Kansas. However, there is no inherent reason why the KOX methodology will not be equally effective in other areas, even applied at vastly different scales or with entirely different variables. The technology exists to evaluate the efficacy of these variables as guides to oil and to assess their spatial reliability in an exploration context. By incorporating the extensive production and exploration data files that have been gathered for the major petroleum provinces of the world, this approach might provide the most realistic assessments of undiscovered reserves that could be made.

REFERENCES CITED

Demirmen, F., 1973a, Numerical description of folded surfaces depicted by contour maps: Jour. Geology, v. 81, p. 599-620.

——1973b, Probabilistic study of oil occurrence based on geologic structure in Stafford County, south-central Kansas: KOX Tech. Rept., Kansas Geol. Survey, 188 p.

Grender, G. C., L. A. Rapoport, and R. G. Segers, 1974, Experiment in quantitative geologic modeling: AAPG Bull., v. 58, p. 488-498.

Griffiths, J. C., 1969, The unit regional-value concept and its application to Kansas: Kansas Geol. Survey Special Dist. Pub. 38, 48 p.

Matheron, G., 1963, Principles of geostatistics: Econ. Geology, v. 58, p. 1246-1266.

Prelat, A., 1974, Statistical estimations of wildcat well outcome probabilities by visual analysis of structure contour maps of Stafford County, Kansas: KOX Tech. Rept., Kansas Geol. Survey, 103 p.

Robinson, J. E., and M. J. Ellis, 1971, Spatial filters and FORTRAN IV program for filtering geologic maps: Geocom Programs, no. 1, 21 p.

——H. A. K. Charlesworth, and M. J. Ellis, 1969, Structural analysis using spatial filtering in interior plains of south-central Alberta: AAPG Bull., v. 53, p. 2341-2367.

A Quantitative Geologic Approach to Prediction of Petroleum Resources [1]

R. W. JONES[2]

ABSTRACT   A satisfactory solution to the problem of estimating petroleum resources in frontier areas is not available and will not be available until substantial numbers of strategically located wells are drilled. In the meantime, we must keep trying to develop better predictive techniques, and this basically means finding ways to maximize the input of our increasing knowledge of the generation, migration, and accumulation of oil and gas.

Some sharpening of estimates of the ultimate productivity of basins and petroleum zones can be made by quantifying the concept that the richness of a basin in reserves per cubic mile is a function of the amount of, and interaction between, four equally important factors: reservoir (R), trap (T), source (S), and migration (M). If these four factors are defined in a dimensionally consistent way, the reserves per cubic mile for a basin (or a sedimentary packet of any size) can be estimated from the formula: Estimated reserves/cu mi = $R \times T \times S \times M$, where estimated reserves per cubic mile are in thousands of barrels and R, T, S, and M are rated on a normalized 0-10 scale. The 10 rating was empirically determined from the maximum value observed in over 50 well-explored basins. In general, both the ease and accuracy of determining R, T, S, and M decrease from T to R to S to M. Thus, the basic reason for the current increased interest in source and migration studies is that they have the greatest potential for increasing the accuracy of prediction.

INTRODUCTION

My remarks are primarily directed at the prediction of petroleum resources in underexplored basins. They do not apply to the important, but distinctly different, problem of extrapolating curves for well-explored basins.

Most approaches to prediction of petroleum resources are analogies based on statistical or geologic reasoning, or some combination thereof. We simply do not have the requisite knowledge about the generation, migration, and accumulation of petroleum to approach reserves prediction from a genetic viewpoint—although much work is being done to overcome this deficiency.

Unfortunately, none of the wide variety of analogical approaches to prediction of petroleum resources is trustworthy because each basin is unique. Thus, we cannot expect to predict accurately the petroleum resources on the relatively unknown continental shelves and slopes by any type of analogy. Not even a geologic similarity of 99 percent between basins is enough to guarantee any similarity in petroleum resources. One crucial difference in geologic parameters can completely negate the effect of all the similarities. Such limitations are simply facts of life in oil exploration. The analogical techniques used by oil companies do not, and cannot, yield the degree of accuracy required by various administrative bodies concerned with the development of the nation's resources.

The biggest imponderable in the energy-supply picture for this country in the next 25 years is the amount and distribution of the recoverable petroleum in Alaska and the continental shelves and slopes of the United

[1]Manuscript received, February 6, 1975.

[2]Standard Oil Company of California, La Habra, California 90631.

Appreciation is expressed to Standard Oil Company of California for permission to publish this paper. I also thank Kent Johnson of Chevron Overseas Petroleum, Inc. for aid in the development of the factor model discussed in this paper, and the many geologists of Standard Oil Company of California who provided the geologic data on which the model is based.

States. Estimates vary widely, and none of the approaches presented and
discussed in this volume is going to do much either to narrow the range
of estimates or place a higher probability on what ultimately will prove
to be the best estimate. The fact is that our precision in estimating
the petroleum resources of Alaska and the offshore is nearly an order of
magnitude less than for other domestic sources of energy. I do not see
how rational planning of the rate of development of alternative energy
sources can be done in the face of this uncertainty. The required accuracy can be obtained only by drilling wells at many strategic locations
in order to obtain factual data for evaluation.

This paper (1) comments on some of the analogical techniques discussed in this volume and elsewhere and (2) offers a type of analogy
that tries to quantify basin comparison, thereby putting maximum pressure on the explorationist to use all his experience and knowledge plus
all the available data.

TECHNIQUES FOR PREDICTING PETROLEUM RESOURCES
BY ANALOGY: OBSERVATIONS

Basin Classification

Whether or not one uses the framework of plate tectonics, some basin
types clearly have a greater average productivity (reserves per cubic
mile) and a greater chance of containing a large field than others. The
major problem with basin classification is that the number of classifications can approach the number of basins. Then one has beautifully described the basins in the present sample, but achieved absolutely no
predictability for the ultimate reserves of the next basin added to the
sample. The alternative approach of using only a few classifications
also yields only a moderate amount of information, primarily because of
the resulting wide spread in reserves per cubic mile within each classification. Thus, if the number of basins in a particular classification
was high, one might reasonably predict an upper limit to the probable
ultimate reserves for a new basin of the same class. However, unless
the productivity of all the basins in the particular classification was
low, one would probably not come close to predicting the ultimate reserves
of any particular basin. For example, the productivity of the predominant "type" of California basins ranges from 0 to 3×10^6 bbl/cu mi.
What would one predict as the most probable productivity for another
basin of the same "type"?

Statistics

Usefulness of the many possible statistical approaches is strongly
limited by the uniqueness of basins. We can determine certain geologic
parameters that correlate with large reserves per cubic mile. However,
when we deal with specific basins, there is the ever-present risk of
overlooking the one or more negative parameters that can completely offset the effect of many favorable ones. Even if we reject the hypothesis
of basin uniqueness, there looms the "bugaboo" of small sample size. In
practice, this means that it is difficult to know whether one is simply
describing properties of a sample, or has developed a predictive tool to
apply to basins outside of the sample. Some of the statistical approaches
have another drawback, which is shared with geologic methods. Both
geologic and statistical "fits" have a very marked tendency to overvalue
the less productive basins and to undervalue grossly the much smaller

TABLE 1.   SOME GEOLOGIC DIFFERENCES BETWEEN ASMARI FOLD BELT,
SOUTHWEST IRAN, AND AREA OF EAST TEXAS FIELD

Southwest Iran	*East Texas Field*
1. Located in an orogenic foredeep	1. Located on flank of broad regional uplift
2. Traps are huge compressional anticlines	2. Trap formed by slight bowing of gentle regional unconformity
3. Reservoir locally dips >50°	3. Reservoir dips uniformly <50 ft per mile
4. Large reservoir thickness (≈1,000 ft or 300 m)	4. Thin reservoir (100 ft or 30 m)
5. Oil column several thousand feet	5. Oil column ≤200 ft (60 m)
6. Very low initial porosity	6. Very high initial porosity
7. Very high fracture permeability	7. No fracture permeability
8. No unconformities	8. Trap at intersection of two unconformities
9. 15-20 similar traps	9. Only one trap
10. Perfect evaporite seal	10. Chalk seal
11. Source rocks obvious	11. Source rocks not conspicuous
12. Short-distance vertical migration	12. Moderate to long-distance lateral migration.

percentage of highly productive basins. Neither statistical "fits" nor
the drawing of optimistic basin analogies is an adequate substitute for
understanding geologic processes. In truth, there is no reason why we
should expect either of them to be.

"Average" Reserves Per Cubic Mile
and Sediment Volume

Some papers in this volume contain justified criticism of the historical approach to estimating petroleum resources by multiplying a
figure for average reserves per cubic mile for well-explored basins by
a less-explored sediment volume under consideration. Suggestions have
been made in this volume and elsewhere which basically attempt to input
some geology in an effort to distinguish the few potentially highly
productive basins or basin segments from the much larger number of basins
with a lower-than-average productivity. This latter effort is clearly
in the right direction, but it is not forceful enough. Much more geology
and much more knowledge of the dynamics of oil generation, migration,
and accumulation can be injected into our attempts at prediction of
petroleum resources. The following section outlines one way of maximizing
the geologic input.

FACTOR MODEL

Introduction

Ideally, we need a prediction model that can incorporate all we know
about the geology of a basin, all the explicit analogies we can find,

TABLE 2.   SOME CRITICAL SIMILARITIES BETWEEN ASMARI FOLD BELT,
SOUTHWEST IRAN, AND AREA OF EAST TEXAS FIELD

---

1.   Large trap(s)

2.   Large volume of reservoir rock in trap position

3.   Adequate source to fill trap(s)

4.   Very effective migration system

---

and all our experience. Add to this all the developing knowledge of the
interaction between geology and the physical and chemical processes that
control the generation, migration, and accumulation of petroleum. Such
requirements are not easily met. For example, inputting geologic parameters into a productivity prediction scheme is very difficult, primarily
because productivities of basins and of basin segments can be distressingly
different despite many geologic similarities, or they can be nearly
identical despite gross geologic differences. This is well illustrated
by comparison of several geologic parameters that characterize two highly
productive basin segments in southwestern Iran and East Texas (Table 1).
Five-billion-barrel fields can result from two very diverse sets of
geologic parameters. Clearly, the problem is to look past large differences in the specifics of the geology and to determine what prerequisites
of large oil accumulations are common to East Texas and Iran (Table 2).
Factors well developed in each are the following:

1.   *Trap*—Good traps mean that a large volume of reservoir space is
in trap position. At the East Texas field, a truncated sedimentary
wedge trapped at an unconformity beneath an argillaceous chalk is gently
bowed; in Iran, large, high-amplitude, compressional anticlines are
capped with several thousand feet of evaporite-bearing beds.

2.   *Reservoir*—A good reservoir means that a large amount of effective reservoir space exists. In East Texas the reservoir space exists
in a thin, highly porous, permeable sand; in Iran, in a thick, highly
fractured, dense limestone.

3.   *Source*—A good source means that a large volume of petroleum
was generated. In East Texas the generation probably occurred in an
overlapping, moderately organic-rich shale of wide regional extent; in
Iran, it occurred in a series of highly organic marlstones underlying
the reservoir rock.

4.   *Migration*—Good migration means that commercial amounts of petroleum have moved from the source rock and arrived at, and stayed in, the
traps. In East Texas there was long-distance lateral migration of
hydrocarbons associated with water movement; in Iran, short-distance
vertical migration of hydrocarbons, probably in a continuous oil phase,
took place. Conspicuous in both areas are the favorable timing of migration and the subsequent preservation of the accumulations substantially
unaltered.

The next steps in developing a descriptive model for well-explored
basins and petroleum zones were (1) to define and evaluate quantitatively
the four factors (reservoir, trap, source, and migration), and (2) to
combine them so as to determine accurately the known reserves per cubic
mile in the given basins or petroleum zones. One boundary condition
exists: if any factor is zero, the reserves per cubic mile is zero.

This restriction suggests a multiplication model. Even with this limitation, a wide number of options are available. The model discussed here is:

$$\text{Reserves/cu mi} = a \times R \times T \times S \times M.$$

The equal sign is possible because of the definitions of the factors as shown in Figure 1. The method for calculating R and T is shown in Figure 2, and Figure 3 shows one possible way of separately defining the S and M factors.

By use of the formula shown in Figure 2, the R factor for the Los Angeles basin would be:

$$R = 0.28 \times \frac{200}{7{,}758} = 0.0073.$$

The R value of 0.0073 for the Los Angeles basin is the highest calculated for an entire basin in a sample of approximately 50 basins. R can be as high as 1 percent of the basin volume for select basin segments.

To determine the T factor in the Los Angeles basin, we use the formula in Figure 2:

$$T = \frac{110 \times .0066}{2{,}800 \times .0073} = 0.036.$$

In the Los Angeles basin, R is slightly less in the traps (0.0066) than in the entire basin (0.0073), because a substantial number of traps occur on the southwest side where the reservoirs are composed predominantly of the distal portions of turbidites. In basins less geologically homogeneous than the productive part of the Los Angeles basin, T can be calculated for specific sedimentary packets and then summed.

Other sets of consistent definitions are clearly possible, but I have found these the clearest conceptually and the most useful in practice. In fact, the formula is simply an extension of prospect evaluation to basin evaluation. Thus, it can be applied to a sedimentary packet of any size. It is most applicable to petroleum zones as defined by Bois (this volume).

Calibration

The factor model was calibrated on well-explored and geologically well-known basins prior to its use as a predictive tool. The model

$$\text{Reserves (M bbl/cu mi)} = f(\text{R, T, S, M})$$
$$= a \times \text{R} \times \text{T} \times \text{S} \times \text{M}$$

where

a = 26×10^6 M bbl/cu mi

R = Fraction of basin which could contain producible petroleum

T = Fraction of R which is in trap position

R × T = Fraction of basin which is trap capacity

S × M = Fraction of trap capacity which contains petroleum

$$= \frac{\text{Producible petroleum in traps}}{\text{Trap capacity}} = \frac{\%\ \text{full}}{100}$$

FIG. 1—Factor model; a is a constant which converts 1 cu mi to barrels. $R \times T$ is fraction of rock volume which would hold producible petroleum if all traps were full. Product of $S \times M$ defines how efficiently source and migration systems in given rock volume exploited available trap capacity. See Figure 2 for definitions of R and T.

R FACTOR

R = Fraction of basin which could hold producible oil

= R (quantity)	$\times$	R (quality)
= Fraction of basin which is reservoir rock (w/cap)	$\times$	Fraction of reservoir rock (w/cap) which could hold producible oil
= Reservoir (w/cap) (fraction of basin vol.)	$\times$	$\dfrac{\text{Bbl/acre-ft (recoverable oil)}}{7{,}758\ \text{Bbl/acre-ft (reservoir rock)}}$

T FACTOR

T = Fraction of R which is in trap position

$$= \frac{\text{Potentially producible reservoir space in traps}}{\text{Potentially producible reservoir space in basin}}$$

$$= \frac{\text{Trap volume (cu/mi)} \times R \text{ (traps)}}{\text{Basin volume (cu/mi)} \times R \text{ (basin)}}$$

FIG. 2—Definition of R and T factors. R depends on both quantity of reservoir rock and its producibility when heavily oil saturated, T depends on both ratio of trap volume to basin volume and ratio of R in trap to R in basin.

formula is an identity for well-explored basins—that is, all entries on both sides of the equation can be determined independently. By use of personal knowledge and information from many discussions and questionnaires, data were assembled on Reserves, R, T, and $S \times M$ for over 75 well-explored basins, basin segments, and petroleum zones. Only a few of these yielded enough information for separate values of S and M.

Figure 4 shows the values of the various factors for the Los Angeles basin. Both R and T in the Los Angeles basin are near the maximum for the analyzed basins, although some basin segments have higher values.

For ease in comparing factor values for different basins, it is possible to distribute the constant a among the various factors in such a way that whole basins have R and T values that range from 0 to 10 and $S \times M$ values from 0 to 100. However, a few basin segments and petroleum zones would have higher R and T values. Thus the formula can be simplified to: Reserves/cu mi = $R \times T \times S \times M$, where R and T usually vary between 0 and 10, $S \times M$ varies from 0 to 100, and reserves/cu mi is measured in thousands of barrels.

Table 3 shows R, T, and $S \times M$ values calculated for some well-explored basins and basin segments. Informative basin comparisons can be made by comparing factor values and analyzing them in terms of the details of the geology in each basin. For example, a major difference in the three California basins at the top of Table 3 (Los Angeles, Carrizo-Cuyama, Salinas) is their source and migration systems ($S \times M$ values). The differences in the $S \times M$ factor can be related in detail to such geologic parameters as trap timing, depth of erosion, and lack of access of many traps to a mature source.

As indicated in Table 3, the Williston basin contrasts sharply with the Los Angeles basin, in large part because it has only 1/1,000 the trap capacity per cubic mile as does the Los Angeles basin. In addition, the source and migration systems of the Williston basin, despite a very poor grade on any absolute scale, were perfectly adequate for the available trap capacity in the Paleozoic part of the section. Virtually all

$$S = \frac{\text{Petroleum in source}}{\text{Trap capacity } (RT)}$$

$$M = \frac{\text{Producible petroleum in traps}}{\text{Petroleum in source}}$$

$$S \times M = \frac{\text{Petroleum in traps}}{\text{Trap capacity}} = \frac{\% \text{ full}}{100}$$

FIG. 3—One possible definition of S and M factors. Definition of R and T requires that $S \times M$ be defined relative to $R \times T$ (trap capacity) in sedimentary packet being evaluated and not in absolute terms. These definitions obviously require modification in basins or petroleum zones where much of generated oil has migrated out or been cracked to gas, or where extensive alteration of petroleum has occurred in reservoirs.

of the mapped, nonflushed trap capacity is full of petroleum. In contrast, the source and migration systems of the Los Angeles basin were not adequate to fill half the available trap capacity. Because migration in the thick, interbedded source-reservoir sequence in the Los Angeles basin has been very effective, the moderate fill-up can only mean that the reserves of the Los Angeles basin reflect a deficiency of source rock relative to available trap capacity. This latter statement is true despite the fact that the sequence of the Los Angeles basin contains a very high percentage of mature, organic-rich source rock relative to other basins.

Application to Underexplored Basins and Petroleum Zones

The preceding is, of course, preliminary to the problem of estimating factor values for use in predicting reserves in underexplored basins. The estimates are made by using all conceivable pertinent information including (1) direct calculation from hard data, (2) analogy with basins whose factors are known, and (3) derivation from the known and deduced geology.

Factor T is usually the easiest and most precise to estimate. That is why so much money is spent on reflection seismograph surveys. Lacking such data, we revert to various forms of basin analogy. The most useful is probably basin classification. Trap density and type show a higher correlation with basin classification than do reserves per cubic mile.

Prediction of R is a direct challenge to the geologist, requiring him to draw on all his experience and knowledge.

$S \times M$ can be estimated by analogy with the file of well-explored basins, and commonly can be checked by independent estimates of S and M.

In many basins, S is now easier to predict than R. We have learned much in the last decade about the dynamics and quantitative evaluation of petroleum generation.

In the research laboratories concerned with the generation, migration, and accumulation of petroleum, M is the major concern. Nevertheless, we should be looking carefully at the geology in those well-explored basins and petroleum zones where reliable R, T, and S data are available. For these, M can be calculated and a catalog compiled of the combinations of geologic parameters that compose good, bad, and mediocre migration systems.

The factor model is well adapted to Monte Carlo simulation. The most probable, maximum, and minimum factor values are assigned and the resulting

curves appropriately summed to yield a probability distribution of ultimate total reserves.

These techniques have been applied to a variety of basins and basin
segments, two of which are the Eel River basin in northern California
and a portion of the Gulf of Alaska (Fig. 5).

Eel River basin—The Eel River basin is a small Tertiary basin lying
onshore along the California coast 200 mi (322 km) north of San Francisco.
For a long time, spasmodic attempts have been made to find major production there. However, the total result has been discovery of a little
gas. The numbers generated by factor analysis suggest that major production is unlikely. The R factor is nearly an order of magnitude less
than in the highly productive California basins, primarily because the
clastic section was derived from Franciscan metasedimentary rocks rather
than from Nevadan granites. Factor T is only slightly less than in the
highly productive California basins, but $S \times M$ is low. Potential source
rock is abundant. However, the basin is only about 10,000 ft (3,050 m)
deep, the geothermal gradient is low, and the organic matter in the
potential source rocks is basically immature. Consequently, significant
generation of liquid hydrocarbons has not occurred. Most of the known
traps contain only a little gas. A projected 12×10^3 bbl/cu mi (oil-
equivalent gas) is optimistic.

Gulf of Alaska—The situation in the Gulf of Alaska between Middleton
Island and Yakutat is entirely different from that at Eel River, both in
geology and in possible importance to the undiscovered petroleum resources
of this country. The Tertiary section is over 25,000 ft (7,600 m) thick,
the area is replete with large compressional anticlines, and oil seeps
are abundant onshore; also, there is one small oil field. The great
unknown is reservoir capacity. T is probably near 20, and it is probable
that the trap capacity ($R \times T$) will be over half full (i.e., $S \times M > 50$),
unless R is much larger than anticipated. Even an R as low as 1 (one
seventh that of the Los Angeles basin) signifies over 1×10^6 bbl/cu mi,
or well over 10 billion bbl of ultimate reserves. However, an R as high
as 1 is rather unlikely, and R could be 0. Unfortunately, there is no
way to estimate rationally where R lies between 0 and 1 except by drilling. No basin analogies are available. This portion of the Gulf of
Alaska, being mainly unlike any other basin in the world, epitomizes the
uniqueness of basins. The geologic history is very complex and not de-
cipherable in detail

Similar evaluations to those presented for the Eel River basin and a
portion of the Gulf of Alaska clearly could have been obtained by many

Reserves (M bbl/cu mi)	=	a	X	R	X	T	X	SM
	=	26×10^6 (M bbl/cu mi)	X	0.0073	X	0.036	X	0.40
	=	2,700 M bbl/cu mi						

FIG. 4—Los Angeles basin described by factor model. $R = 0.0073$ means that 0.73 percent of basin
volume is reservoir space physically capable of containing producible petroleum. $T = 0.036$ means that
3.6 percent of volume defined by R is in trap position. $S \times M = 0.40$ indicates that 40 percent of potentially productive reservoir space in trap position contains producible petroleum. R, T, and $S \times M$ were
calculated using detailed geologic data provided by Standard Oil of California geologists working in Los
Angeles basin.

TABLE 3. USE OF 0-10 (Apx) RATING SCALE IN ANALYSIS OF VARIATIONS IN RESERVES PER CUBIC MILE[1]

Selected Basins	1 R (Fraction of Basin)	2 T (Fraction of R)	3 R (Rating Scale)	4 T (Rating Scale)	5 S X M (% Full)	RTSM (Col. 3 X 4 X 5) Res./Cu Mi (M Bbl)
Los Angeles	0.0073	0.036	7.3	9.3	40	2,700
Carrizo-Cuyama	0.0068	0.024	6.8	6.3	10	430
Salinas	0.0062	0.046	6.2	12.0	5	380
Big Horn	0.0042	0.009	4.2	2.3	10	100
Powder River	0.0039	0.003	3.9	0.8	10	30
Portion Powder River Cretaceous	0.0016	0.006	1.6	1.6	40	100
Michigan	0.0009	0.002	0.9	0.6	30	16
Paradox	0.0009	0.005	0.9	1.3	15	18
Williston (U.S.)	0.0008	0.0003	0.8	0.1	95	8
Iran (Dezful embayment)	0.0008	0.15	0.8	39.0	90	2,800

[1]Constant a (26×10^6) in factor model was proportioned among various factor values as follows:

R was multiplied by 10^3,
T was multiplied by 2.6×10^2, and
$S \times M$ was multiplied by 10^2.

Basin	R	$\times$	T	$\times$	$S{\times}M$	$=$	$RTSM$ (M bbl/cu mi)
Eel River	0.8	$\times$	3.0	$\times$	5	$=$	12
Portion of Gulf of Alaska	0–1	$\times$	20	$\times$	50	$=$	0–10^3

FIG. 5—Application of factor model to Eel River basin and portion of Gulf of Alaska between Middleton Island and Yakutat.

other techniques. I prefer the factor model because it requires the geologist to make use of his total experience and knowledge.

SUMMARY AND CONCLUSIONS

Each basin is unique, and our knowledge of the dynamics of petroleum migration and accumulation is woefully inadequate. As a consequence, no technique exists, or can be expected soon to exist, for predicting petroleum resources in relatively unexplored basins in the United States that can approach the accuracy required by the various administrative bodies concerned with the development of the nation's energy resources. The certainty of failure does not remove the responsibility of trying, however.

From the geologist's viewpoint, the problem is how to maximize the input of all the available geologic data, all his experience, and all the rapidly developing knowledge of the interaction between geology and the physical and chemical processes that control the generation, migration, and accumulation of petroleum.

The factor approach discussed in this paper is only one of many possible ways to make the best of what is basically an untenable situation from the standpoint of national planning. More reliable estimates are needed, especially of Alaskan and offshore reserves, in order to permit planning of the rate of development of alternative energy sources. The necessary knowledge regarding Alaskan and offshore petroleum resources can be obtained only by drilling many wells in strategic locations and thereby obtaining factual data.

REFERENCE CITED

Bois, Christian, 1975, Petroleum-zone concept and the similarity analysis—contribution to resource appraisal: AAPG Studies in Geology 1 (this volume).

Index

This index consists of two sections, which appear in the following order:
(1) Author Index and
(2) Keyword Index.

The *author index* is arranged alphabetically according to each author's last name. For papers by more than one author, each author's name appears in the index in alphabetical order. The appearance of an author's name followed by the title of an article does not mean that he is the only author of that article. He may be one of two or more authors of the paper whose title follows his name. The author index does *not* show multiple authors in any single listing.

To locate a reference in the *keyword index*, the reader should begin by thinking of the significant words. Then he should look in the index for the keyword entry for each of those words. The reference codes will direct him to the pages.

The columns on the right-hand side of the keyword index give the page number (1 or 3) indicating the nature of the source. The code is:
(1) for phrase from title; and
(3) for phrase from abstract, text, table, figure, or figure caption.

The keyword for each entry is located at the left-hand side of the page. The ($>$) sign indicates the first word in each title or key phrase. The ($<$) sign indicates the end of the title or key phrase.